MYTH AND ENVIRONMENTALISM

This volume traces the interconnections between myth, environmentalism, narrative, poetry, comics, and innovative artistic practice, using this as a framework through which to examine strategies for repairing our unhealthy relationship with the planet. Challenging late capitalist modes encouraging mindless consumption and the degradation of human–nature relations, this collection advocates a re-evaluation of the ethical relation to "living with" and sharing the Earth.

Myth and the environment have shared a rich common cultural history traveling as far back as the times of storytelling and legend, with the environment often the central theme. Following a robust introduction, the book is organized into three main sections—Myth, Disaster, and Present-Day Views on Ecological Damage; Indigenous and Afro-diasporic Myths and Ecological Knowledge; Art Practices, Myth, and Environmental Resilience—and concludes with a Coda from Jeanette Hart-Mann. The methodology draws from diverse perspectives, such as ecocriticism, new materialism, and Anthropocene studies, offering a truly interdisciplinary discussion that reflects on the dialogue among environment and myth, and a broad range of contributions are included from Canada, the United States, the Caribbean, Ukraine, Japan, Morocco, and Brazil. The book joins a long line of approaches on the interrelations between ecological and mythical thinking and criticism that goes back to the early 20th century.

This volume will be of interest to students, scholars, activists, and experts in environmental humanities, myth and myth criticism, literature and art on more-than human and nature interaction, ecocriticism, environmental activism, and climate change.

Esther Sánchez-Pardo is Professor of English at Complutense University, Madrid, Spain.

María Porras Sánchez is an Assistant Professor at Complutense University, Madrid, Spain, and a literary translator.

ROUTLEDGE EXPLORATIONS IN ENVIRONMENTAL STUDIES

Series Editors
Charlotte Burck and Gwyn Daniel

Urban Aerial Pesticide Spraying Campaigns
Government Disinformation, Industry Profits, and Public Harm
Manuel Vallée

Addicted to Growth
Societal Therapy for a Sustainable Wellbeing Future
Robert Costanza

Coastal Disaster Risk Management in Bangladesh
Vulnerability and Resilience
Edited by Mahbuba Nasreen, Khondoker Mokaddem Hossain and Mohammed Moniruzzaman Khan

A History of Radioecology
Patrick C. Kangas

Satire, Humor, and Environmental Crises
Massih Zekavat and Tabea Scheel

UN Human Rights Institutions and the Environment
Synergies, Challenges, Trajectories
Sumudu Atapattu

Myth and Environmentalism
Arts of Resilience for a Damaged Planet
Edited by Esther Sánchez-Pardo and María Porras Sánchez

For more information about this series, please visit: www.routledge.com/Routledge-Explorations-in-Environmental-Studies/book-series/REES

MYTH AND ENVIRONMENTALISM

Arts of Resilience for a Damaged Planet

Edited by Esther Sánchez-Pardo and María Porras Sánchez

Routledge
Taylor & Francis Group
LONDON AND NEW YORK

earthscan
from Routledge

Designed cover image: Trace Hudson © iStock

First published 2024
by Routledge
4 Park Square, Milton Park, Abingdon, Oxon OX14 4RN

and by Routledge
605 Third Avenue, New York, NY 10158

Routledge is an imprint of the Taylor & Francis Group, an informa business

British Library Cataloguing-in-Publication Data
A catalogue record for this book is available from the British Library

ISBN: 978-1-032-39135-9 (hbk)
ISBN: 978-1-032-39134-2 (pbk)
ISBN: 978-1-003-34853-5 (ebk)

DOI: 10.4324/9781003348535

Typeset in Times New Roman
by Taylor & Francis Books

CONTENTS

FIGURES

CONTRIBUTORS

Jeanette Hart-Mann is a farmer, artist, activist, and teacher whose research and practice is rooted in agroecology, environmental justice, and eco-social storytelling. She is faculty at the University of New Mexico where she teaches art & ecology through field-based programming. Hart-Mann co-founded and co-directs SeedBroadcast and HawkMoth Farm Agroecology Center.

Paul Humphrey is Associate Professor of LGBTQ Studies and Africana & Latin American Studies at Colgate University, NY. His research focuses on gender, sexuality, and spiritual practices in Caribbean cultural production. His monograph, *Santería, Vodou and Resistance in Caribbean Literature: Daughters of the Spirits*, was published with Legenda in 2019.

Hope Jennings is a Professor of English at Wright State University–Dayton Campus. Her areas of research and teaching include women's writing, ecocriticism, posthumanism, and apocalyptic literature. Recent essays appear in *Tulsa Studies in Women's Literature, Contemporary Women's Writing, LIT: Literature Interpretation Theory*, and *ISLE: Interdisciplinary Studies in Literature and Environment.*

Christine Junker is a Professor of English at Wright State University–Lake Campus. Her teaching and research areas include literature and the environment, women's literature, and the scholarship of teaching and learning, with essays appearing in *ISLE: Interdisciplinary Studies in Literature and Environment, College Teaching*, and *Studies in the Humanities.*

Haley Laurila completed her PhD in Slavic languages and literatures at the University of Michigan where she wrote her dissertation on Chornobyl. Her research focuses on the Soviet nuclear legacy and memories of environmental disaster. She is currently a graduate student at Wayne State University and is working in the digital humanities.

Esther Lezra is an Associate Professor of global studies at UC–Santa Barbara. Her research focuses on the literary and cultural study of the Caribbean, Europe, and North Africa, 18th to 20th centuries. She has published in *Dissidences, Anthurium, Journal of American Studies*, and *Kalfou*. Her book, *The Colonial Art of Demonizing Others*, was published by Routledge in 2014.

Leonor María Martínez Serrano is a Lecturer at the University of Córdoba, Spain. Her research interests include Canadian literature, ecocriticism, and high modernism. She has co-edited *Modern Ecopoetry: Reading the Palimpsest of a More-than-Human World* (Brill, 2021) and authored *Breathing Earth: The Polyphonic Lyric of Robert Bringhurst* (Peter Lang, 2021).

María Porras Sánchez is an Assistant Professor at Complutense University, Madrid, and a literary translator. She has co-edited, with Esther Sánchez-Pardo and Rosa Burillo, *Women Poets and Myth in the 20th and 21st Centuries* (Cambridge Scholars, 2018) and with Gerardo Vilches, *Precarious Youth in Contemporary Graphic Narratives* (Routledge, 2022).

Sonja Ross is an independent scholar. She holds a PhD in ethnology, and her research focuses on mythologies, rituals, worldviews, and cultural change in the Circumpolar Zone, North/South America, and Eurasia. Between 2000 and 2012 she was the first chairwoman of the *Circle of Friends* at the Museum of Ethnology in Munich.

Esther Sánchez-Pardo is Professor of English at Complutense University, Madrid, Spain. She is the author of *Cultures of the Death Drive* (Duke UP, 2003) and *Antología Poética, Mina Loy* (2009) and the editor of seven books, recently including *Poéticas Comparadas de Mujeres* (Brill, 2022).

Lhoussain Simour is Associate Professor of English and cultural studies at Hassan II University of Casablanca–Morocco and Senior Research Associate at University of Gibraltar. He is the author of three books, including *The Construction of Marginalities and Narrative Imaginary in Mohamed Zafzaf's Texts* (Lexington Books, 2022).

Keijiro Suga is a poet and Professor of critical theory at Meiji University, Tokyo. Formerly the president of ASLE–Japan, he has written on animal lives, rewilding, and ecocriticism. His many translations include works by such writers as Edouard Glissant, Maryse Condé, and J.M.G. Le Clézio.

ACKNOWLEDGEMENTS

We are most grateful to the writers and artists (and copyright holders) who have generously granted us permission to reproduce their work: Greg Anderson-Elysée, Hugo Canuto, Jeanette Hart-Mann, shalan joudry, Mohamed Larbi Rahhali, Edgardo Miranda-Rodriguez, and Maki Ohkojima.

We also wish to gratefully acknowledge the excellent editorial team at Routledge. Without the wise advice, expertise, and insight of Grace Harrison, Editor for Environment, Sustainability and Product Design; the generosity and expertise of Matt Shobbrook, Senior Editorial Assistant for Environment and Sustainability; and the knowledgeable suggestions of three anonymous reviewers, this project would probably not have come to fruition.

The Editors' work in this volume has been funded by Comunidad de Madrid (Aglaya H2019/HUM/S714, with contributions from the European Social Fund).

INTRODUCTION

Myth and environmentalism: entanglements, synergies, openings

Esther Sánchez-Pardo

Introduction

The present volume aims to address the many challenges posed since the second half of the 20th century by cultural discourses at the interface between culture and science within the field of myth and environmental studies. In the territories of myth and environmentalism, this volume attempts to bridge the gap between the contradictions of a late capitalist society in its ambivalent encounter with a bountiful nature, provider of goods and well-being, and an exhausted and overexploited nature on the verge of collapse. Few human constructions are more "social" than myth. In other words, without community or social group, myth would never exist. In its turn, the environment, under its many guises, adds up to the idea of the "natural," the aggregate of social and cultural conditions that influence the life of individuals and communities on the planet (Merriam-Webster 2022). Thus, sociality and environmentality[1] are intertwined areas where being and knowing closely interact within a web of interdependence and transformation.

Myth and the environment are not rigidly bounded fields; rather, they are expanded areas of experience, practice, and knowledge growing out of a socially constructed understanding of nature—"It appears that history or culture rests on nature, when in fact the reverse is true; nature becomes, in effect, a social creation [...]" (Evernden 1992, 24)—the social, and the human and more-than-human worlds.[2] Received ideas on nature and different notions of the "natural" through time make us aware of its social and historical construction. Social constructivist theories postulate that the definitions of "nature" and "environment" differ according to culture and time, without denying the existence of an external nature or environment, strictly speaking, but rather emphasizing that there is only a diversity of conceptions influenced by the social, political, and cultural contexts.

DOI: 10.4324/9781003348535-1

According to the *Oxford English Dictionary* (2022), the meaning of environment has changed since the beginning of the 20th century. Environment then usually referred to "an area surrounding a place or thing." In the 1940s, the term started to acquire the meaning we now associate with environmental thinking: "The natural world or physical surroundings in general, either as a whole or within a particular geographical area, especially as affected by human activity." In identifying the difficult interrelation between humans and their environments, a disproportionate unequal relation based on careless exploitation and predatory usurpation of resources, this volume will focus on the current state of unbalance and its consequences.

Postmodernism provoked public debate about the collapse of grand narratives, the provisional nature of truth, and the cultural construction of knowledge, and no discipline reduced its huge impact. It subverted foundational beliefs in identity, history, and objective knowledge. It brought to the fore awareness of how imperialism, cultural appropriation, and discrimination (gender, race, sexuality, religion, ableism) divided the world between the victors and the vanquished. Within this bleak picture, the importance of ideology as a mediator between fact, subjectivity, and sociality, together with rampant inequality and the depletion of the world's resources, erected insurmountable barriers between peoples and put the whole world at risk.

The aftermath of postmodernism—as post-postmodernism, alter modernism, under its many labels—is still in progress. Nevertheless, this is an era marked by critical debates on connectivity and mass communications, transnational social movements, data science, algorithmic bias controversies, and the digitalization of life, where we find ourselves exposed to environmental hazards and climate change.

The resurgence of interest in myths and their social and cultural value in our time draws strength from the feeling that the interrelation between humanistic and scientific and technological realities is difficult to understand fully and to incorporate as meaningful domains of human experience. Zygmunt Bauman (Bauman and Donskis 2015) has warned us about the loss of community in an increasingly individualistic world, where individuals "struggle to invest sense and purpose in their lives [...]" (Bauman 2001, 13). In his view, this is the major task of sociology. He has argued that progress is a myth (Bauman and de Querol 2016), because people no longer believe that the future will improve their living conditions. For Bauman, we are in a period of interregnum in the midst of substantial transformations between a time of certitudes and one of uncertainties.

Within this context, myth continues to be a particular kind of narrative. It is a scaffolding, a structure that sustains social life. Folklore studies scholar Elliot Oring defined narrative as "another word for story." For Oring,

> Narrating is a method by which an experience is transformed into a verbal account. [...] It should be noted that nowhere in this definition is

> myth held to be untrue—rather, that the narrative is held by someone to be ultimately true enables its characterization by the folklorist as myth.
>
> *(1986, 121, 124)*

Although in Western culture "myth" and "fiction" are nowadays used almost interchangeably,

> the *mythos/logos* contrast, like the one between imagination and reason, fictive and factual, was established in the fifth century, along with mythology, and linked to critical inquiry in both history and philosophy in ways that correspond clearly to common understandings of 'myth' and 'reason' in the Western tradition.
>
> *(Fowler 2011, 66)*

Nature and reality are mirrored in myth, and the frame, the substance, and the major elements—setting, conflict, resolution—of myth "appear to be products of 'nature' rather than history—expressions of a transhistorical consciousness or some form of 'Natural Law'" (Slotkin 1992, 6).

Environmental and ecological[3] concerns have long been present in literature and the arts, but the urgency and crisis of the present moment demand that we take action now. Fiction and art practices raise awareness and become a rallying cry in support of conservationism, sustainability, and reparation to regain livable conditions for the whole planet. Traditional accounts of myth have enhanced our relations and understanding of the environment. The anthropological, philosophical, and sociological study of myth, together with the scientific evaluation of climate change damage, deserve further investigation to grasp the interactions between science and culture in a continuum rather than as worlds apart.

To better confront the difficulties that climate change, toxic waste, the proliferation of microplastics and their impact on the whole planet pose, it is urgent that we strive for a reparative environmental justice through education. By providing information and using a more interdisciplinary approach to understanding and interpreting the many signs of environmental exhaustion, we raise awareness on how crucial an ethics of sustainability and environmental education are for re-establishing a healthy inter and intra-active relation with all co-participants of life on Earth.

Crucial as well is Donna Haraway's (1988) perspective of "situated knowledges," which acknowledges the diversity of knowledge systems and challenges the notion of a unique scientific truth. Haraway maintains that all knowledges are the product of power relations and are shaped by those very relations. Thus, all knowledges represent a partial perspective and help constitute social location, but "the ability partially to translate knowledges among very different—and power differentiated—communities" remains a project worth pursuing (580). Environmental thinking works, using Lorraine

Code's words, with a material and "situated subjectivity for which place, embodied location and discursive interdependence are conditions for the possibility of knowledge and action" (2006, 20).

Finally, the notion of "environmental literacy" is at the core of this volume as well: "[I]t comprises an awareness of and concern about the environment and its associated problems, as well as the knowledge, skills, and motivations to work toward solutions of current problems and the prevention of new ones" (McBride et al. 2013, 3). Environmental literacy is a must in critical times, and using it to teach for environmental justice mobilizes cognition and affect. As Bruno Latour argued, "A common world, if there is going to be one, is something we will have to build, tooth and nail, together" (2004, 455). This specific literacy is more relevant at present than it has ever been, and scholars claim that it is precisely environmental health that plays a major role in promoting the broad goals of environmental literacy. When people make the connection between health and the environment and are totally aware of the impact of environmental hazards on themselves, environmental literacy achieves success (Chepesiuk 2007).

In the chapters that follow, reading, comparing, and reflecting upon the current permanence of myth in a global context fraught with environmental problems and challenges, across geographies and cultures, will allow us to appreciate and assess what comes out of the interaction and adaptation of mythical and environmental modes of knowledge. Our interest lies in cultural work that critically responds to the challenges human action has posed for the environment and that imagines alternatives to the current situation, one so difficult to reverse.

Reading for myth and environmentalism in this volume

This volume aims to make it easier to understand the cultural implications that myth in its entanglement with the environment brings about in terms of their reciprocal interactions and in their exchanges, influences, intersections, and derivations. The social, political, esthetic, and ethical implications of the mythical environmental and of the environment that gained mythical status will be addressed within the web of significance (Geertz 1973) that connects them to their rich specific context and larger transnational framework. One of the major questions addressed in this volume is to what extent environmental literature and art register the problem of calamity and catastrophe and how their forms embody the various tensions and contradictions inherent to the conceptualization of our current ecological crisis. These tensions bring to the fore the difficult conceptual knots at the center of so many environmental questions. Thus, for example, environmental critics have only just begun to discuss how those at risk of displacement by climate change and its effects may actually end up subjected to the precarious new environments where they would seem to be relocated, usually under the protection of a rich Western country.

As aptly stated by Nathan Jandl (2013), environmentalism gains urgency precisely due to the ominous affective resonance of "urgency" itself in a complex and highly unbalanced global scenario. We are caught in a temporal trap between the grim present world and the disintegrating future planet we are bequeathing to later generations. Environmental humanists and communities paying tribute to the ritualistic and mythical aspects holding them together remain very much attuned to the societal, scientific, and symbolic transformations necessary to achieve change. Art and literature become especially important to explore new possibilities and to help humans imagine alternatives to the current state of affairs (Jandl 2013). In the meantime, a broader social debate continues about how ideas on green politics, sustainability, renewable energy, and a livable life are in dire need of a radical change, which usually goes with a profound transformation of the many neoliberal regimes that shape the global economy.

Environmental activists, scientists, and analysts have sounded the alarm with what they call the "sixth extinction" (Kolbert 2014) but only modest agreements to limit emissions from coal- and fossil fuel–dependent economies—emerging and rich countries—are reached in climate summits.[4] Even if scientists urge world leaders to prevent global temperatures from rising above 1.5°C—because warming above this point could unleash irreversible and uncontrollable climate impacts—industrial activities, extractive practices, and aggressive chemicals continue to be a source of pressure on the environment.

The increase in atmospheric gas concentration from fossil fuel emissions has had an enormous impact on climate warming and, associated to it, sea level rise (Siegert 2016). Oceans and water masses are progressively warming and are being filled with plastics and other pollutants to a substantial level, the consequences of which are under study and have yet to be seriously evaluated. Huge areas of natural forests, including the Amazon rainforest, are being destroyed in favor of farmland and logging operations, with an increased monetary value. This has resulted in a decrease in carbon dioxide uptake and habitat destruction. The swift pace of habitat reduction is currently the leading cause of biodiversity loss. Global warming is a direct consequence of the industrial emission of greenhouse gases, and this has led to a measurable surface air temperature increase of 1°C since 1880 and a sea level rise of 20 cm over the same period (IPCC 2013).

Environmental decay and damage are usually circumscribed to the two poles around which public debate seems to be organized: the idea that change is intrinsic in nature and therefore unavoidable and the apocalyptic discourse preempting the end. Despite the seeming irreconcilability of these views, neither can be totally disregarded. In our view, myth and environmental literature and art both narrate environmental crisis and offer solutions through the employment of ecological scenarios where human advancement converges with nonhuman preservation. In the following chapters, strategies for the conceptualization of nature and human and nonhuman interaction will be discussed.

Another major question we seek to address is how society contributes and reacts to manifestations of sustained critique, community building, and networks of transnational concern in both eco-art and literature. People across a range of geographical locations, cultures, and economic conditions request solutions to the enormous challenges of species extinction, climate change, differentially distributed pollution and toxification, and the politics that propel and reinforce it (Jandl 2013).

Communities, held together through a shared history, group identity, networks of solidarity, myth, and rituals, are also attached through a sense of place to local landscape and common habits. Many of these communities see their surroundings radically altered by economic forces and a deregulated, unrestricted use of natural resources. Grassroots activism most often pivots around the feelings of attachment to a local landscape, destroyed by aggressive practices such as mining, power plants, megadams, and toxic dumping, among others (Jandl 2013). Nevertheless, collective responses tend to remain localized.

Protest is mostly channeled through nongovernmental organizations (Greenpeace, Earth Justice, WWF) and social movements. Individual communities badly affected by environmental damage also stand up for the right to enjoy a clean, healthy, and sustainable environment. At present, social and environmental justice movements and indigenous rights movements contribute to the struggle over land and resources. Due to the heterogeneity of these groups and organizations, scholars are raising important questions that must still wait to be properly addressed. For instance, Chris Crews calls for an "Earthbound cosmopolitics." In his words,

> What sets the Earthbound cosmopolitics apart from existing movements for justice is the explicit call for an expanded view of politics that beckons people into active defense of the land and is rooted in an Earth-centered philosophical and ethical framework.
>
> *(Crews 2019, 350)*

This volume will critically examine whether alternatives to the current political situation favoring neoliberal economies and extractive capitalist practices are not only envisioned but also enacted. Recent initiatives, such as those promoted by Fridays for Future, Extinction Rebellion (originally UK based, now worldwide), the climate movement, Animal Rebellion, and many other nonprofit grassroots organizations, aim at countering the dramatic, long-lasting effects that nonsustainable exploitative practices have on our planet.

Throughout this volume, we envision eco- and environmentally friendly projects within the trajectory we contemplate that runs through epistemologically rich initiatives. In Lorraine Code's authoritative view, "Ecological thinking examines the contributions of epistemic and ethico-political practices to producing habitats where people can live well together and respectfully with and in the physical natural world" (2006, 279). These epistemologies, grounded in horizontal practices respectful of differences, as Code writes, "combine careful readings of differences characteristic of empiricism in its creative, deliberative versions with investigations that locate events, experiences, symptoms, social issues, problems, within wider patterns of power and privilege, oppression and victimization, scarcity and plenty, joy and sorrow" (280).

As pointed out by Allison Athens (2013), projects in environmental humanities include care for diverse and unevenly enfranchised human groups, other species, and the ecosystems that should allow the survival of all plant, animal, and microbial others. Other than critical attention to the human that unites most of these projects is a concern for ecological crisis (Athens 2013). Ideas and responses to crises will be presented and discussed in subsequent chapters of this book.

Finally, while addressing questions of long-term societal resilience, how can we approach climate change across the disciplines and what happens if we fail to do it?

The notion of resilience has circulated widely in fields like ecology and psychology for some time. Just recently, over the last 15 years, it has gained prominence within politics and security where previous technologies have failed to control nontraditional security threats such as war, financial crises, and environmental catastrophe.

Resilience is a complex concept that "offers a theory of growth, development and improvement through embracing change, diversity, surprise and disruption, rather than banishing these conditions beyond the limits of the sovereign subject" (Chandler et al. 2020, 3). In "resilience thinking" one must "grapple[s] with the challenge of how to ground truth claims in the face of indeterminacy" (6). For our purposes, it is interesting to note that resilience approaches seek to disrupt the divide between human/nature or subject/object, because it is not only humanity that enjoys agency, and resilience resituates both agency and creativity also on the side of nature/the nonhuman world.

The capacity of millions of people to build resilience to climate and disaster risks, mitigate climate change, and cope with loss and damage must be enhanced and implemented through community cooperative models. It is thus crucial to analyze and understand the intersection between climate justice and gender, class, and ethnic justice. This proves critical to inspiring social actors to leading the way toward more democratic and inclusive social and environmental approaches. Finally, resilience focuses on relations over entities, and when one speaks of environmental resilience, the reference is to systems whose transformation and changes affect a wealth of elements and variables (Chandler et al. 2020).

The planetary turn

We owe to Gayatri Spivak some of the most crucial ideas on the planet and the planetary turn in recent times.[5] Situated within comparative literature,

Spivak calls for the discipline to reinvent itself within the domain of "planetarity." In her view, the mobility of populations worldwide, crossing borders and inhabiting new and alien spaces, would benefit from claiming the planet instead of the globe or the world to imagine themselves. Both the globe and the world are easily manageable. The Earth and its inhabitants fall under the aegis of globalization, which means under the primacy of the neoliberal order. The planet in itself is a vast and, to a great extent, unknown entity, "The planet is in the species of alterity, belonging to another system; and yet we inhabit it, on loan" (Spivak 2003, 72).

In Spivak's view, there is always an ethical component in the encounter with the other—"To be human is to be intended toward the other" (2003, 73)—and with more radical or less radical entities, "[M]other, nation, god, nature. These are names of alterity, some more radical than others. Planet-thought opens up to embrace an inexhaustible taxonomy of such names [...]" (73). The planet, in its incommensurability and radical otherness, is far from being possessed or mastered by any regime. Its uncanny qualities—Spivak evokes Feud's *unheimlich*, as translated into English by the Stracheys—are ever present; they pervade our sense of place, our home. Its problematic representability allows us, in Spivak's examples, "to construct an allegory of reading [...] by way of nationalist colonialism and postcoloniality" (73). Spivak's rich display of "planetary" readings illustrates well how we are increasingly dealing with heterogeneity on a scale vaster than ever before, always in our journey toward human and nonhuman others. In Spivak's work, planetarity proves to be a new paradigm in progress.

Also within the literary studies domain, the "planetary turn" has been an ever-growing trend in other areas and chronologies. Susan Friedman, scholar in modernist studies, alerted us to the importance of taking the whole planet as a unit of analysis. The spatiotemporal dimensions of imagining humanity's dwelling space and historical lived time within a "planetary" scale "invok[ing] the Earth in deep time,"[6] Friedman states,

> The Earth moves in a scale of time almost unimaginable in human terms, although we attempt to name its ages and periodize its changes, often cataclysmic ones. How might we conceptualize modernity anew in the context of the Earth's and many of its species' indifference to the human? Conversely, what are the effects of human modernities on the Earth and the nonhuman, a question that discussions of rapid climate change and the Anthropocene attempt to address?
>
> *(2018, 78)*

In Friedman's view, planetarity should be used as an epistemology, not an ontology.

Prominent historian Dipesh Chakrabarty has elaborated on the planetary turn and written on the material and intellectual implications of polar ice cap melting on peoples in the South (Chakrabarty 2009). In his latest book, *The Climate of History in a Planetary Age* (2021), Chakrabarty sets out to investigate how historians in particular and human scientists in general should confront the undeniable reality of climate change, the result of human technological intervention on Earth on a planetary scale. Chakrabarty places us in the Anthropocene, an age where humanity has managed to erase the borders between human history and natural history. Because, through technology, humans are acting on Earth and altering geological time, Chakrabarty posits that we are beyond a "global" age and well into a "planetary" age and that history, as all human and social sciences, should redefine itself to account for this paradigm shift, where the Earth rather than the human now takes center stage. Although he holds humans as a whole responsible for the Anthropocene, he coins the concepts of retrospective guilt, to be applied to richer economies long responsible for damage to the environment, as opposed to prospective guilt, for those which only recently joined in force. Without denying the major role played by the former economies in the onset of the Anthropocene, he rightly points out that "the scientists' discovery of the fact that human beings have in the process become a geological agent points to a shared catastrophe that we have fallen into" (Chakrabarty 2009, 218–219).

Throughout his latest book, Chakrabarty provides examples of how we should adjust and reappraise our thinking to Earth times rather than human times. Paramount to this reappraisal is our redefinition of political thought, a political thought that has traditionally remained "indifferent to the biosphere" (Chakrabarty 2021, 211). No specific solution is given for this. In the conclusion—in the form of a dialogue between Bruno Latour and Chakrabarty—both thinkers agree to the use of "terrestrial" as a way to address a new type of necessary subjectivity that also needs to "resignify the word 'subject'" (211).

Clearly, we should admit our own complicity in resource exhaustion and industry, companies, and businesses their responsibility in resource exploitation. Human groups, communities, cities, and countries are inextricable from a long human history of resource exploitation. Humans now, no matter how much they stand and struggle to preserve the environment, embody the contradiction that they continue to deplete its resources to maintain a modern standard of living along with the industrial economy upon which that standard is predicated. By invoking epochal changes that might reverse this status quo, Chakrabarty hopes that humanity realizes "that the socioeconomic-technological arrangements we currently have cannot go on indefinitely" (2021, 212). Technology modifies the environment on a planetary scale.

Myth in times of (environmental) crisis

The contemporary social crisis is one of cultural meaning-making (Kuipers 2019) where rituals have become devoid of their traditional power to unite people by mutual agreement, shared activities, and a joint focus on shared symbols (Alexander 2004).

How can we remedy the aforementioned? Sociologist Michele Lamont has recently argued that we need to create and sustain new cultural narratives. In her view, there is an "urgency of developing a more sophisticated understanding of recognition gaps and how to address them. [...] This can be achieved in part by promoting new narratives of hope via scripts of self that broaden cultural membership" (Lamont 2019, 667).[7] Lamont argues this can be achieved through "agents of change" in education, philanthropy, religion, advocacy, and other sectors, contributing directly or indirectly to diffusing new frames about our collective future and identities. What is it that myth might offer as a narrative of hope, a bond that gives cohesion to our sense of community? The enduring power of myth shows in its resilience and its malleability. Myth remains open to change and transformation, subject to re-creation; its adaptability to different times and its prescient narrative semanticity make this discursive mode into one of the most successful in the history of literature and ideas. Myth has gained the attention of disciplines such as anthropology, philosophy, folklore, religious studies, art, and literary history.

From a philosophically solipsistic position, the human individual would never confront such a construction. Myth is a sociohistorical construction; it is born in the midst of a society looking at itself in a mirror where origins, tradition, *idées reçues*, consensus, mores and habits, and co-existence hold. Antagonism and dissent are also one more sign of the emergence of the social forces at work, as in the opposition between rural and urban populations or within the mythical scenario, the frequent and patent antagonism between parents and offspring. These common elements are significantly enhanced by a series of narratives that bring cohesion and help to hold the social fabric together. When placed within an expanded global map of exchange and interaction, the purported harmonious coloration of myth can easily be disrupted by making radical differences apparent and incompatible with a peaceful consensus.

Myth making can be understood as an ongoing creative process that involves daily interpretations and potential additions, in which people put together fragments of "meaning-laden narratives" into ever changing constructions that constitute variations on well-known stories (Zelnik 2016, 229). These microfragments of myth circulate widely, and individuals take those they need to help them make sense of challenges difficult to cope with. Popular culture is a crucial component of myth, and in its current shape, in the age of the internet, it has become less locally specific and more widespread.

Conscious of the fact that there are many different kinds of myth—creation myths, social myths, myths about the physical world—hero myths count among the most prominent—this volume will cover myth and environmentalism as they interact and produce new knowledge or finally modify what society, scholarly work, and the current state of the transmission of myths on the environment hold on to within a transversal global culture that is attentive to differences. Due to the very broad scope of both areas, studies on myth and environmentalism must necessarily be partial and noncomprehensive.

The turn to myth and the environment in the 20th and 21st centuries

We can trace the contours of a refurbished theory on myth and the environment starting in the mid-20th century and chronologically arranged. The itinerary we propose reconsiders major contributions to myth as always entwined with the environmental and vice versa. Major critics and researchers like Roland Barthes, Richard Slotkin, and Elizabeth and Paul Barber and writers like Paula Gunn Allen and Tomson Highway have significantly problematized, enriched, and expanded the massive archive of knowledge on myth and the environment.

Roland Barthes' *Mythologies*, originally a collection of monthly articles that appeared between 1954 and 1956 in *Les Lettres Nouvelles*, was published in 1957. Many of the essays deal with current events and others with broader aspects of French popular culture. These two aspects, the everyday and the broader dimension of culture converge in modern myth, pervade the fleeting phenomena of the "quotidien" with a repertoire of stereotypes and *idées reçues*. One of the major features of modern myth is the transformation of historical phenomena into timeless essences. Myth is a temporal anomaly that arrests the flow of history. It operates to substitute historical consciousness by the alleviating comfort of a timeless present. Barthes, as a fine semiologist and cultural critic, sets out to analyze the strategies of popular mythic representations that endow the contingent with an essential quality. These are the means by which mass culture remakes history and transforms those elements into ideology.

Barthes devotes one of the sections in *Mythologies* to plastic. His acute analysis shows this special awareness that places him always ahead of his time. Plastic, in its long history since the turn of the 20th century, is an element that necessarily takes us to current debates on environmental damage and climate change. The French critic speaks of plastic as the manifestation of a "sudden transformation of nature" (Barthes 1972, 97). One of its major utilitarian advantages is resistance, and it is a clear sign of the "evolution of the myth of 'imitation' materials" (98). These imitation materials belong in the domain of the bourgeoisie, to the world of "appearances" (98). Nevertheless, plastic has also become a household material and thus prosaic, common, and not rare. Plastic has come to modify the "age-old function of nature," an artificial element that is pervasive, and "is about to replace [nature]" (98). In Barthes' view, plastic performs an "equalizing function" by virtue of which "the hierarchy of substances is abolished" (99). Clearly, plastic exerts this eerie attraction of the "unknown" (in its origin, in its composition), and it may replace the actual, exclusive and elitist commodified objects, with its "quick-change artistry" (97). As Jennifer Wagner-Lawlor writes,

> Obscured by the symbolic economy of a "plasticized" (195) world of consumers, buoyed by innovation and so-called convenience, Barthes sees a toxic underside: a world of users with an instrumentalist attitude toward other human beings, toward nature, toward life itself, it is not just the aesthetics of plastic that Barthes rejects.
>
> *(2017, 97–98)*

If traditional materials, those regarded by many as natural, retain a sense of their earthly origin, plastic is unique in retaining nothing of its raw materials' earthiness. Its artificiality bears witness to the technologization of consumption and everyday life. The innovations that late 20th-century technology brought about promised to liberate humanity from the predicament of its biological past.

Barthes ([1957] 1972) elaborates on myths as socially, historically, and politically determined. Myths are never innocent, and they transmit a particular ideology. Through semiotic analysis we can identify two levels of signification: a first-order linguistic system that Barthes calls the "language-object" (114),—alluding to the literal meaning and to linguistic denotation—and a second-order system that transmits the myth and he names a "meta-language" (114)—speaks about the first order and exists in the domain of connotation. Within this second order of connotation, the signs take on the value of the culture in which they are embedded.

In Barthes' extended essay in the section "Myth Today," he displays his semiotic model in order to explain how cultural texts (logos, advertisements, proverbs, …) can transmit several levels of meaning. In his view, myths remove context and, thus, crucial historical information from signs, turning them into "an empty parasitical form" (Barthes 1972, 117). Society, culture, and science move on quickly and, for Barthes, progress is conceived either in terms of scientific materialism or within the human sciences. "Myth, conversely, remains subject to regression and stasis. […] Science proceeds rapidly on its way, but the collective representations do not follow, they are centuries behind, kept stagnant in their errors by power, the press, and the values of order" (1997, 37–38). Myth is unprogressive, even when attached to novelty, and the tension that exists between ostensive advancements and ailing representations pervades all the critiques in Barthes' *Mythologies*. The latest fads that go with new consumer products get dragged down with deep-seated beliefs and mystifications.

Historian Richard Slotkin has devoted effort to reflect upon social myths. Slotkin is an authority in studies on the US civil war and has approached the relation between foundational narratives and nature in his extensive research on the myth of the frontier. In his view, the latter, with its violent conquest of indigenous people and landscape, has been at the origin of American national identity. Myth and ideology walk side by side in Slotkin's theorization. In an attempt to offer a critical interpretation of the myth of the frontier, he assesses "its power in shaping the life, thought and politics" of the United States from 1890 to the present (1992, 4). In his view,

> Over time, through frequent retellings and deployments as a source of interpretive metaphors, the original mythic story is increasingly conventionalized and abstracted until it is reduced to a deeply encoded and resonant set of symbols, "icons", "keywords" or historical clichés. In this form, myth becomes a basic constituent of linguistic meaning and of the processes of both personal and social "remembering." Each of these mythic icons is in effect a poetic construction of tremendous economy and compression and a mnemonic device capable of evoking a complex system of historical associations by a single image or phrase.
>
> *(5–6)*

In Slotkin's view, myth-making processes "lie in our capacity to make and use metaphors, by which we attempt to interpret a new and surprising experience or phenomenon by noting its resemblance to some remembered thing or happening" (1992, 6). In his words, if the metaphor proves appropriate, individuals will be inclined to treat the new phenomenon as a reverberation of the old. A myth is generated through an applicable metaphor that accurately explains a current reality in association with a past myth. The myth-making process is complete when a metaphor offers a cogent diachronic explanation among the past, present and future. When reality is translated into myth, "the complexities of social and historical experiences are simplified and compressed" (13).

As Slotkin (1973, 1992) amply demonstrated, a mission to transform the savage wilderness into an organized civilization underlies the frontier myth. The myth is invested with a "pastoral impulse" (Kolodny 1975, 6) that attempts to bridge the gap between the growing accretion of memory over time and reality. Leo Marx argued that "the imagination of Americans was dominated by the idea of transforming the wild heartland into such a new 'Garden of the World'" (1964, 141), so that "the quality of their feeling [...] presumably expressed a believable definition of reality" (143). If, with Kolodny, we agree that this impulse is able to "shape and structure experience" (1975, 6), we can infer from here that many unexpected relationships arise between myth, reality, and the pastoral, all central to the myth of the frontier. This pastoral impulse becomes the major catalyst driving myth into reality, transforming a chaotic wilderness into an orderly universe, ruled by Old World values. As Slotkin has demonstrated, this "errand into the wilderness" allowed white Americans to justify their "imperialistic adventure[s]" (1973, 562) and to impose their ethos on the new land. Thus, it has become commonplace that the frontier figure (a white male) managed to tame the chaotic wilderness of the New World into a "civilized" garden.

As Anette Kolodny rightly argued, "Eden, Paradise, the Golden Age, and the idyllic garden, in short, all the backdrops for European literary pastoral, were subsumed in the image of an America" (1975, 6). This crucial element in the US imaginary has become a founding myth that reverberates in the history of the nation. Along this diachronic line, Slotkin has argued that myth proliferates in times of crisis and, interestingly enough, "in the end, as the historical experience of crisis is memorialized and abstracted, the revised ideology acquires its own mythology, typically blending old formulas with new ideas or concerns" (1992, 6). Slotkin alludes to how in moments of disruption or popular discontent an attempt is made to revise the underlying intellectual and moral content of ideology and the received wisdom embodied in myth. This approach to the transformation of (social) myth emphasizes historical contingency and the activity of human culture producers (anonymous authors) in modifying new forms of the cultural expression conveyed by myth.

Elizabeth and Paul Barber (2004) have recently attempted to find explanations for mythological stories based on natural phenomena or real events. They have given credence to highly controversial stories by providing interpretations of given myths based on what they claim as evidence. What interests us at this point is the fact that the majority of their reports on how to read those myths are based on the direct observation and study of the natural world. In their view, a myth's function is to transmit crucial information about the natural world. In a very ambitious project drawing from archaeology, linguistics, folklore studies, and cognitive science insights, their endeavor has contributed to revitalizing debates long overdue on the relation of the "new" humanities (environmental, urban, global) for the 21st century and the sciences (neuro- and cognitive sciences, sustainability, computer sciences). One of the leading theses in their work states that myth served as a medium of recollection in a time when writing was not yet available. People used the myth as a mnemonic device, preserving essential data and endowing them with a suitable form. Due to the limitations of human memory, they put those data into "the oral pipeline for distant descendants" (160) for transgenerational communication.

In their work, they enumerate a large number of "myth principles" and attempt to propose some kind of mythological hermeneutics that teaches readers a series of "decoding" procedures to understand myth. The process is called "the stripping procedure" and it assumes that accretions to whatever happened originally have sedimented over time. Multiple layers of analogical, semantic, and syntactic operations by virtue of what they call the restructuring principle have produced cultural change. This change leads, in due time, to reinterpreting the data that myths strive to convey. The mythic–linguistic principles they refer to can be reduced to the functions of silence, analogy, compression, and restructuring and to a series of subprinciples that specify these functions.

In Elizabeth and Paul Barber's view, myths were never intended as fictional accounts; rather, they were carriers of important cultural information over very long historical periods. Most of their examples have to do with volcanism or refer to astral "data." Thus, for example, early on in their volume, they discuss Crater Lake (Oregon), a lake formed around 7,700 years ago by the collapse of the Mount Mazama volcano. The Klamath (Native Americans) regarded the lake as a sacred place and the abode of the "great Spirit," and it has remained so up until today. In the authors' view, "[R]eal information can reach us intact across more than seven millennia of retelling" (2004, 8).

Myths encode observations and events, but the rationale that clarifies what renders these materials into myths is never laid out in the Barbers' volume. In any event, the valuable link between myth, language, and cognition is reinforced within an environmental context. The Barbers urge that myths are storages of significant data that, in many cases, can be verified by geology and astronomy. Myths act as constructs that encapsulate observation and knowledge and contribute to make sense of evolution and historical experiences.

In the 1980s and 1990s, within the context of minority literatures, myth became a contested terrain in debates between ethnic essentialism and the poststructuralist (de)construction of knowledge and identity categories. When myth is taken as a foundational yet flexible form of narrative or motif in the literary and artistic contexts, its ability to both make claims to truth while exposing those truths as socially and politically determined is a feature that distinguishes myth's "work of culture."[8] In this regard, myth can be endorsed as a fruitful means to ground a culturally specific literary and art theory with comparative potential (Schiff 2010). In Sarah E. Schiff's view,

> [...] because myth is a communal narrative that bounds a group of people through intersubjective experiences of telling, listening, and reading, and not by the color of their skin or the deity they worship, it has the capacity to evade both racial essentialism and complete deconstruction while providing a tangible object of study: a narrative.
>
> *(293)*

For her, myths do not make reference to

> "first-order" things in the way of history, yet they also do not make reference to what exists purely in the individual imagination in the way of fiction. Myth, in other words, hovers in the liminal generic space between history [...] and fiction.
>
> *(2010, 151)*

In the context of minorities, myth, used purposefully to inscribe ideas, values, and customs within the social fabric, also works to counter hegemonic historical discourse. When literature and art contain myths from multiple traditions, they bring to light locally specific cultural values and worldviews. By deciphering these myths within their contexts, one can fully appreciate which of them make conservative appeals to tradition and which challenge hegemonic views that have remained unchanged throughout history to the detriment of vulnerable groups.

For Native American Laguna-Pueblo writer Paula Gunn Allen, myths belong to their community. She claims,

> American Indian myths depend for their magic on relationship and participation. Detached, analytical, distanced observation of myth will not allow the listener mythopoeic vision. Consequently, these myths cannot be understood more than peripherally by the adding-machine mind; for when a myth is removed from its special and necessary context, it is no longer myth; it is a dead or dying curiosity. It is akin, in that state, to the postcard depictions of American Indian people that abound in the southwestern United States.
>
> *(2015, n.p.)*

In her work, she has aligned the scholarly employment of myth with other derogatory terms that imply backwardness and general mockery of the ways in which the natives understand the world. Allen has argued that in such a context, myth has always been used to question accuracy and legitimate knowledge:

> Popularly among Americans, myth is synonymous with lie; moreover, it implies ignorance or a malicious intent to defraud. Thus, any attitude or idea that does not conform to contemporary western descriptions of reality is termed myth, signifying falsehood. Labelling something a myth merely discredits the perceptual system and world-view of those who are not in accord with the dominating paradigm. Thus, current dictionary definitions of myth reinforce a bias that enables the current paradigm of our technocratic social science-biased society to prevail over tribal or poetic views just as it enables an earlier Christian biblical paradigm to prevail over the pagan one.
>
> *(2015, n.p.)*

As such, myth stands opposed to truth whenever it is considered as a part of a fiction–fact binary, a conundrum that many anthropologists and social scientists find themselves unable to resolve.

Just to illustrate the controversy around the appropriation of the myths originating in minorities or vulnerable populations, in her scholarly work, Allen has critiqued Leslie Marmon Silko for publishing stories that, for Allen, belong to the tribal community. One could argue that by recording traditional myths in her novels and by departing from their exact telling, Silko is revising the role of myth and both making it available to her community and opening it up for other readers. Thus, within the use of myth in her text, Silko exposes the conventions of narrative emplotment[9] at work and shows that myth can be used to grant visibility and voice to those silenced due to colonial impositions and extractive capitalism.

Renowned Cree author Tomson Highway defines myth as "a collection of stories that record the spiritual culture across time and space" (2003, 19). In Highway's view, mythology establishes cultural solidarity by giving symbolic form to the experiences, both positive and negative, shared by a specific group of people (19). In his work, Highway compares and reflects upon similarities and differences among Christian, Greek, and Aboriginal mythology, with a special focus on his Cree heritage. He highlights the importance of myth for First Nations people:

> I am here to talk about mythology because I believe that without mythology, we would be nothing but walking corpses, zombies, mere empty hulks of animal flesh and bone, skin and blood and liquid matter with no purpose, no reason for existing, no use, no point, nothing, mere flesh and bone and skin and blood with nowhere to go, and with no guide to guide it through a life path that, one imagines, has been given to us all by … what? Who? Why? And why here? These are the questions mythology answers.
>
> *(18)*

In his creative work, Highway uses the principles of traditional mythology to weave stories from contemporary experience. In fact, he has made a conscious effort to call upon Greek mythology and use it in a syncretic operation with Cree and other native mythologies. In Highway's work, aboriginal mythology stands out as a major area of interest and expertise, focusing on "the role of Aboriginal mythology within the context of world mythology. In particular, its pivotal importance to the issue of environmental health, preservation and survival" (Highway 2022). Highway explains that for the Cree the universe arose from O-ma-ma, "a miraculous entity eventually to be known in the English language as Mother Earth" (Highway 2003, 39). In the Cree context, this force is female, but in the wider context of native mythology it is both male and female simultaneously (41). And most important, "God [is] in everything […] we speak of 'pantheism'" (42). Central to First Nations' spirituality is the idea that all beings in nature are endowed with qualities that make them subjects of respect and of rights, "from leaves to soil to water" (42). In Highway's account of existence, the universe is an endless "circle of death and life and re-birth and death and life and re-birth" and all those who lived prior to us remain with us still today, "in the very air we breathe, in the shimmer of a leaf on that old oak tree, in that slant of sunlight that falls in through your window and lands on your wrist" (44). Finally, comparing mythologies, when it comes to the notion of paradise or the Garden of Eden, there is no such myth either in aboriginal or in Greek mythology. Highway pays tribute to the First Nations' land—and by extension, of North America, "the most spectacularly beautiful continent on earth" (46)—clearly "a gift from a benevolent female god" (46) where the Tree of Knowledge remains, "right there in the middle of the garden for us to partake of, for us to enjoy, for us to celebrate every day, twice thrice a day, if we have to" (47). In Highway's view, neither science nor religion have ever been able to explain exactly where the impulse of the first cell in the universe came from. A new language had to be invented to articulate that origin, and that language is mythology. In brief, for Highway, "mythology [lies at] the exact halfway point between science and religion, truth and fiction [...] that most elaborate of all fictions" (49).

Once again, we will move from myth to environmentalism in the following section, where we will provide a reflection on how late capitalist production places the planet under extreme pressure. Economics, as a major social discipline dealing with scarce resources (water, land, human life) the mere shortage of which causes their market value to rise, works as a heuristic approach to analyze inequality and comparative indexes of development. At this point, ecological economics, a new field dating back to the 1980s, teaches the world how sustainable development, based on principles of justice, care, equitable distribution of resources, and preservation of all life on the planet, helps us understand how value should not be simply attributed to material accumulation and explores in what ways nature and the material can coexist in balanced interaction within specific ecosystems.

A view from ecological economics on human (so far) and planet exploitation

From the vantage point of the ecological economy, the notions of economy, production, wealth, and labor have been analyzed ever since the economy emerged as a discipline linked to the origins of capitalism. At first, the discussion revolved around two positions. According to early physiocratic tenets, the only source of wealth was land, and production was related to agriculture. This perspective links the notion of production to materiality and, therefore, to the limits of the biosphere.

For Marxism, the only factor that brings about production is labor through the transformation of natural resources. Capital appropriates such labor in the form of surplus value of the resulting wealth and exploits the workers. In this scenario, the epitome of production is industry, and progress and development are understood, then, as a gradual and progressive industrialization. Here, nature is but a mere resource for progress. The aforementioned implies a first denial of the limits of human action, but at least there continues to be a link with the materiality of the world. This link is severed with the imposition of the ideas of neoclassical economics according to which the only factor that generates production is capital, which uses inputs (natural resources and labor) to produce new value, namely, monetary value. In brief, wealth is money and production consists in the generation of monetary flows. Finally, development is measured in the constant increase in labor productivity. The progressive displacement of human labor by technology is an outgrowth of this state of affairs. Natural resources remain hidden under the form of monetary value. Naredo clearly explains this conceptual shift toward understanding capital (read, money) as the hegemonic factor of production:

> Capital could substitute for Land and labour, which thus appeared as the ultimate limiting factor in the production of wealth. A conceptual closure of the notion of economic system in the universe of pecuniary values is completed [...] once the umbilical cord that originally linked the economic to the physical and human dimensions was cut [...] social concern was deriving from the production of wealth to the mere acquisition of monetary value.
>
> *(2006, 165–166, author's translation)*

By setting production apart from natural resources and the human factor, and by considering that the economy is that which involves an exchange value, the possibility of infinite growth appears. There is only one way for the anthropocentric fantasy of production to continue: to get a phantom medium with which to accumulate that wealth we are supposed to create; this is achieved by turning money into a means of accumulation. This facilitates the appearance of wealth generation as a process that can continue indefinitely and allows the problem of limits be circumvented. Finally, it sets economic activity apart (that is, monetized activity) from any link with the material and energy resources that support it.

In any event, these limits remain in place. This ever-growing production requires an escalating level of consumption of natural resources and energy, and it also entails a rate of waste emission that is unsustainable. In Martinez-Alier's view,

> [O]ne must see the economy as composed of three levels, not only the financial economy (that got out of hand, and that now wants to impose "debtocracy" everywhere), and not only the real, productive economy; but also the real-real economy of the flows of energy and materials and the unavoidable production of waste.
>
> *(Martinez-Alier and Pellegrini 2012, 351)*

All new production leads to the depredation of natural wealth. Wealth generation appears as a fact, but this is unreal because it is based on the disappearance of natural wealth. This trespassing of the limits is not distributed in an equitable way at all; rather, it is inextricably linked to global inequality, because what is unsustainable in itself are the modes of production and consumption that go with development. There is a profound intergenerational (we live off the planet that belongs to future generations) as well as intragenerational inequality (some parts of the world live at the expense of the part of the planet that belongs to others). The current critique of production and overexploitation amounts to the critique of an anthropocentric fantasy that entails environmental depredation and global social injustice.

Naredo (2006) explains that the "metaphor of production" has taken over our way of interpreting the world with the erroneous conviction in our ability to produce wealth as an endless process with no limits. This has become the major socioeconomic goal: progress and development understood as endless growth. The planet, the set of natural resources available for man to dominate in order to "create civilization," remains subservient to his demands. Nonetheless, all this is an anthropocentric fantasy, arising from an understanding of the world that places human activity at the center of all processes. The metaphor of production is governed by an anthropocentric logic that denies eco-dependence by setting human life apart from the rest of life on the planet and by opposing culture to nature. The latter is a mere repository for the human creation of wealth (by means of either capital or human labor).

The feminist critique

The ongoing climate crisis has helped us to better perceive the huge gaps and discrimination in terms of gender, race, class, ableism, and precarious legal status, which undermine the lives of millions of people. Feminist economist Amaia Pérez Orozco reflects on this conflict in her 2014 book *Feminist Subversion of the Economy. Contributions for a Debate on the Capital–Life Conflict*, which addresses why climate activists and feminists agree in denouncing a system that has instrumentalized and endangered both the natural world and people, especially those who are oppressed because of their gender, class, race, or geographical location.

Women are presented on the one hand as victims of environmental degradation who need to be rescued and, on the other hand, as heroines leading initiatives to halt the destruction of nature. Simultaneously, they are burdened with responsibility for family and subsistence work, discrimination based on gender being instrumentalized as part of the solution. Clearly, this essentialist rhetoric toward women—especially toward poor and racialized women from the Global South—ignores the strategy hidden behind the intended link between women and nature, further excluding men from the latter and associating them with culture. Women and vulnerable populations are, indeed, the most affected by climate change and natural disasters. In the specific case of women, their vulnerability is not innate but the result of inequalities produced by social gender roles, racism, classism, and patriarchy that translate into social norms, cultural practices, and laws that are discriminatory and that finally reproduce poverty.

The feminist approach delves into the structural causes of inequality, studying roles and responsibilities and how these are attributed to a given gender. It assesses the differentiated climate risks faced by women and men in all their diversity, LGBTQ and other identities, of all ages, beliefs, classes, and ethnicities. Based on this understanding, we gather that feminist climate policies are resilient policies in the face of climate change. Feminist ecocritic Val Plumwood has proposed a critical ecological feminism that can:

> [...] provide a basis for a connected and co-operative political practice for liberation movements. We need a common, integrated framework for the critique of both human domination and the domination of nature—integrating nature as a fourth category of analysis into the framework of an extended feminist theory which employs a race, class and gender analysis [... feminism is a] vital contribution to a more complete understanding of domination and colonization.
>
> *(1993, 1–2)*

In brief, we are not all equal when it comes to experiencing environmental damage and lack of climate justice. There is no gender, class, or ethnic neutrality when it comes to addressing sustainability, and there will be no climate justice without a feminist and intersectional approach.

Environmentalism and postcoloniality

As the field of environmentalism and ecocriticism evolves, many have agreed on the need to integrate other regions of the world—outside the West—and also to incorporate the postcolonial situation in the reading of the environment. It is to this end that the last decade has seen the proliferation of research reconciling postcolonialism and ecocriticism, a way of correcting the Eurocentrism of the movement: *Postcolonial Ecocriticism* (2010) by Graham Huggan and Helen Tiffin; *Postcolonial Green* (2011), edited by Bonnie Roos and Alex Hunt; *Postcolonial Ecologies* (2011) by Elizabeth DeLoughrey and George Handley; and *Environment at the Margins* (2011) edited by Byron Caminero-Santangelo and Garth Myers. There is a clear desire to integrate formerly colonized regions, although many of these texts are limited by the lack of analysis of works from other-than-English-speaking areas.

The majority of these texts mention Rob Nixon's article, "Environmentalism and Postcolonialism" (2005), as the first essay to address provocative elements regarding the reconciliation of postcolonialism and ecocriticism. Rob Nixon indeed shows in four points the levels of exclusion between the two theories in these terms: "Postcolonialists emphasize hybridity, while ecocritics emphasize Purity; postcolonialists study displacement, ecocritics focus on place; postcolonialists tend toward cosmopolitanism, ecocritics toward nationalism; postcolonialists work to recover history, ecocritics seek to sublimate or transcend history" (235). Despite the points of divergence that seem to definitively separate the two methodologies, ecocriticism and postcolonialism manage to find connecting bridges. At least that is what the authors of recently published books on postcolonial ecocriticism do.

Following the ideas of Huggan and Tiffin, Roos and Hunt believe that the main point of convergence between the two methodologies is the concept of justice: "In our view, it is the term 'Justice' that provides a space for theoretical work bridging and merging ecocriticism and postcolonialism" (2011, 3). To this affirmation one could add that ecocriticism and postcolonialism are linked at the level of their position of marginality. Without reproducing the pattern of domination they denounce, postcolonialism and ecocriticism strive to put oppressed cultures and dominated nature back at the center, respectively. Bonnie Roos and Alex Hunt reconcile the two methodologies by assigning them common objectives. They claim:

> As we see it, postcolonial green scholarship must define itself not as a narrow theoretical discourse but as a relatively inclusive methodological framework that is responsive to ongoing political and ecological problems and to diverse kinds of texts. No particular approach or methodology can fit all cultural contexts or geographies, but any text, we suggest, can profitably be read from a postcolonial green perspective.
>
> *(9)*

While noting the interdisciplinary character of postcolonial ecocriticism, Roos and Hunt also underline the inclusive property of this new discipline, which deconstructs the politics of exclusion of formerly colonized cultures and of nature in relation to culture. Thanks to this integration, the different regions of the world can participate in the environmental discourse by bringing in their specificities. This is what Roos and Hunt explain by saying:

> In our view, scholars certainly must account for the fact that what it means to be "postcolonial" or "green" varies radically in different geographies. But it is our contention in this collection that difference should not unduly hamper our ability to listen and to learn from each other. Scholars must be able to draw comparisons in order to further a theoretical, postcolonial, green project—in order to learn methods of resisting injustice in its various forms.
>
> *(2011, 7)*

It is by taking into account the uniqueness of vast geographical areas like Africa, Southeast Asia, and Latin America that we propose in this volume several readings on myth and the environment rooted in the African context (East and North), Japan, Eastern Europe, and (First Nations) Canada, among other locations, that suit the realities of the fruitful co-occurrence of myth and environment. The ideas and theories that will be developed and discussed in this volume belong mostly to myth studies and environmental/ecocritical studies, from the critical humanities to current art practices. They all aim to address the interdependence between the cultural and the natural. Thus, for example, cultural imperialism, theorized by Edward Said in *Culture and Imperialism* (1993), allows us to see how imperialism has marked large areas of the world in the environment and local populations. Cultural imperialism has also favored the standardization, the homogenization of the culture and the environment of millions of people. This brings to mind the colonial policies of assimilation that contributed to destroying both cultural and ecological diversity in the colonies through the establishment of monocultures.[10]

At present, and as part of the legacy of postcolonial thinking in its ongoing critique, we should not be fooled into believing that the environment has to do with nature and it belongs to the world outside. The environment, and the planet at large, is an expanded field overlapping politics and social thought, and the time is ripe for decolonizing it.

Environmental art/eco-art: How to do things with living and inert matter

In that realm of experience where perception, cognition, and action interact and the sensuous, emotional, imaginative, affective, communicative, and symbolic coalesce and act as permeable boundaries between inside and outside, art provides space for experimentation and creative exchange. Art has been used to raise awareness of environmental damage, of the dangers of pollution, overexploitation of resources, and climate change. In recent times, environmental art posed a challenge to the postmodern ethos that proclaimed the museum was in ruins (Crimp 1993)[11] but still remained dormant to the hierarchies between artistic disciplines (more established art practices versus performance or ephemeral modes) and life outside the gallery and the "profession." Only the land art of the late 1960s and 1970s involved gestures that can be read as precursors of eco- and sustainable art two decades later.

The critical concerns raised by Rachel Carson's *Silent Spring* (1962) led to the rise of the early global environmental movement. Within this climate of opinion, many of the pioneers of eco-art grew aware of a wild consumer culture paradoxically tolerant of toxic waste, polluting agents, and environmental hazards. Early environmental movements were spearheaded by artists such as Joseph Beuys, Ant Farm, Mario Merz, and Mierle Laderman. Their creative projects aimed at raising awareness of the risks and dangers of environmental degradation for the whole planet. At present, taking action in art is an even more pressing concern; "environmental art 'helps improve our relationship with the natural world'" (Hull 2007).

In Linda Weintraub's view, ecologists and eco artists "[…] actively strive to ensure the vitality of Earth's ecosystems" (2012, 6). In the recent history of the eco-art movement, Weintraub points to the important shift from anthropocentric to ecocentric perspectives, "Ecocentrism envisions humans as components of interconnected systems. [...] These eco-art pioneers challenged the assumptions that nonhuman forms of life are important only to the extent that they are useful to humans" (6–9). Progressively, the ecocentric alternative has been privileged, focusing on interdependence and mutuality, devoting efforts to enhance all life forms, from living organisms to inert forces and all extant entities.

The ecological concerns of the 1960s were an inspiration for many interdisciplinary artists willing to reach out to the environmental domain. This eco-art movement gained momentum in the 1960s and 1970s when, for the first time, environmentalists seriously pondered the damaging effects of human intervention on the planet. Since then, many artists have cultivated an ecologically sensitive art that reflects upon environmental decay and climate change. It was during this period that scientists began to research climate change widely. Carson's *Silent Spring* (1962) was the first essay to specifically deal with global environmental issues. Carson's appeal to all audiences denounced "man's war against nature" (7) and made people aware of the moral obligation to preserve all life forms and their biodiversity: "Most of us walk unseeing through the world, unaware alike of its beauties, its wonders, and the strange and sometimes terrible intensity of the lives that are being lived about us" (249).

Artists today live in a period of climate change because the material conditions and narrative that shape art made at the present time are very much conditioned by these circumstances. As we have examined thus far, the inequalities that go with the huge impact of climate change are expressively marked by the presence of myth and the narratives that reinvigorate it as such. How is myth rearticulated in 21st-century art practices? How does it reappear, and what are the main challenges it poses?

At present, we are in need of what Donna Haraway invokes as "art science activisms [...] working with beauty and fury in enlisting the possibility of ongoingness" (2014). This call for interdisciplinarity by a leading historian of science and human–animal relations scholar aims at opening up debates on foreseeable futures and sustainable living conditions for all on the planet. Her latest research represents opposition to capitalism's relentless expansion and enduring damage. The interdisciplinarity she calls for, so pervasive in current art practices and projects, is in itself a "banner" (Weintraub 2012, Preface) of environmental art.

Artists and critics advocate for a repositioning of the museum and the gallery as open and participatory sites where art takes place. Art appears as integrated with the environment and all living organisms. As artists Gilles Bruni and Marc Babarit argue, "[T]he attitude we advocate, this idea that humanity is part of nature and that our position stands on 'terra firma.' [...] Explicitly we share the already ancient idea of a museum without walls, that the work of art is the site, and that the site is planetary. The art is thus where it is made, with the people who are there" (2004, 26). This expansion of the potential site where art is made, where it occurs and the impact it produces, together with the relational and interactive scenario it brings about, and the entangled web of interactions it is embedded in is, in itself, a consequence of a self-reflective move in which art and artists are co-participants in sustainable futurity with the planet at large.

Finally, countless questions arise: does nature have a will of its own we cannot control (Dougherty 2004)? What is the role myth plays in registering the current state of the planet? Does art need to have an environmental mission at present? Is art working to reveal a specific site, to reconcile a landscape with its dwellers and their mythopoetic abilities? To give back meaning to the local? Or is it rather accompanying and acting as a life-enhancing force that is generative and ever-expanding? How does myth contribute to all these issues?

More on myth and environmentalism in this volume

In an attempt to tackle these and other questions, our volume joins a long line of approaches on the interrelations between ecological and mythical thinking and criticism that goes back to the early 20th century. The following chapters show how both research and creative approaches concerning myth and the environment, in their interconnections and entanglements, have increased very rapidly since the mid-20th century. Providing a global perspective has been a major objective within this project. The Americas, the Caribbean, Africa, Asia, and Europe have taken the lead in teaching us more about how humans, the more-than-human world, and nature engage with myth—both preserving and producing new myth coeval with their societies' evolution and progress—while coping with the current stage of environmental decay.

Whereas the strength of myth lies in its diversity, it is a slippery concept that resists definition. Myth-making narratives in this volume take many forms: (fictional) storytelling; transactional knowledge; cohesive narration; the "quotidian"; shape-shifters of social reality; mediums of recollection (prior to writing); long-term cultural information carriers; cognitive devices between mind, reality, and language; communal narratives; counterhegemonic minority discourses and carriers of spiritual culture across time and space. The forms of myth-making narratives mentioned here are among the most salient. This lengthy list of notions associated to myth can barely be based on any cultural or critical consensus besides their singularities in their historically specific contexts.

This volume is designed as a shared conversation around myth and environmentalism with our readers and any interested party. How does the community of the many—engaged in stopping and studying climate change—

benefit from such a conversation? Activists, scientists, scholars, citizens, migrant populations, women, the new generations, government officials, non-governmental organizations, politicians, and the new acknowledged realities (juridical status granted to landscapes, areas of high environmental value, animals, other actors) should intervene in this urgent, last-minute assembly where every movement and decision are interconnected. Our current assembly has different profiles including activist protesters' sites, Fridays for Future demonstrations, climate summits, grassroots organizations, meetings reclaiming indigenous myth and storytelling traditions to combat the erasure of the natives, citizens' assemblies, and popular initiatives to detoxify our common everyday realities.

We would very much like this volume to help open up a democratic assembly for dialogue, debate, interconnection, and exchange at a global level. Utopian as it may seem, the fluidity of exchange among us—among social network movements and regular citizens—should be presided by a sustainable platform of ideas in which any party might intervene on an egalitarian basis to express, denounce, weave a story, or speak out for the preservation of the planet. By recourse to myth and to all current initiatives aiming at improving the living conditions of humans, the more-than-human and all processual changes that contribute to containing and enhancing cultures, we salute any serious creative enterprise joining efforts in this project.

For the greater good, all actors within the assembly should pay attention to such engaged conversation. The ethics of listening is strongly at work in the following chapters. An active listening and reading are thus required for engaging with the ideas presented in this volume. Recycling old concepts—never wasted or discarded to promote sustainable living horizons—is also a fundamental goal throughout this book.

The exploitation of the planet under the current, very late stage of capitalism produces irreversible changes and severe alterations in the lived experience of present world populations. Our volume is divided into three main sections that correspond to major areas responding to the urgency of this volume's call to action.

In part I, Myth, disaster, and present-day views on ecological damage, apocalyptic rhetoric and images intertwine as much with events around the Chornobyl nuclear catastrophe as they do with the bleak panorama portrayed by extreme narratives of the Anthropocene. Both chapters emphasize myth, ecological disaster, possibilities for reparation, and restorative justice with a focus on ethical responses for a damaged planet.

The European perspective, which is present in chapter 1 ("The afterlife of Chornobyl"), is now probably more present than ever with the war in Ukraine. In our view, it is crucial to reflect upon the Chornobyl catastrophe from a European and, specifically, Eastern European perspective. An insider's view focusing on the work of two artists, Markiyan Kamish and Fedor Aleksandryvych, is a much-needed piece for reflection.

In “The afterlife of Chornobyl: apocalyptic mythology and environmentalism in the Exclusion Zone,” Haley Laurila explores the relationship between mythologization and nuclear power in the case of the Chornobyl nuclear disaster. She examines the disaster’s multifaceted mythological dimensions, from the Soviet myth of nuclear utopia that led to the construction of the Chornobyl power plant to the apocalyptic mythology describing its collapse, as found in the destructive psychological and material traces of disaster. Her chapter focuses specifically on the Chornobyl Exclusion Zone, which, despite its reputation as an abandoned wasteland, has become a nature reserve and an example of postdisaster renewal of both plant and animal life. As an apocalyptic space, the Zone attests to not only the environmental traumas of nuclear power but also the potential for change embodied by the tentative renewal of nature that has occurred there in the absence of humans. From this perspective, Chornobyl’s apocalyptic dimensions are not oriented toward a finality but actually reveal opportunities to change the ways we think about the future of the planet and our responsibility to build a sustainable environmental future. Stalkers and other illegal visitors to the Zone are uniquely positioned to explore the terrain and map these changes. Using the experiences of writer Markiyan Kamysh and artist Fedor Oleksandryvych as a foundation, the author discusses how the knowledge cultivated by proximity to disaster can help make the long-term effects of our destructive exploitation of the environment visible. Apocalyptic thinking can help us understand Chornobyl’s long-lasting impacts on the environment, as well as the specific historical contexts and ideologies that sustain ecological harm. In the context of the Anthropocene and our increasing nuclear precarity, this kind of local knowledge of the Zone can inform the more ethical and expansive environmentalism we need to embrace in order to prevent further ecological breakdown.

Within the panorama of recent work on the Anthropocene, this chapter connects nicely with “Myths of wilderness and motherhood in post-apocalyptic narratives of the Anthropocene,” in which Hope Jennings and Christine Junker respond persuasively to myth by addressing the idea of myths of wilderness and motherhood as a social construction in the imaginary of the United States. Their reading of narratives of the Anthropocene is framed within a feminist perspective, finding default in the equation that operates when comparing the current understanding of nature and wilderness with motherhood. The underpinnings of mythological motherhood are, in the authors’ view, the desire to believe that there is some external entity, whether it is a mother or an environment, dedicated to or shaped just around our own desires, along the frustration and grief of realizing that no such entity exists. In this interesting co-relation between mythical motherhood and the environment, Jennings and Junker examine the anxiety and fears produced by living in a world on the edge of human extinction. At this juncture, survival is the only thing that counts, and without a sustainable, ethical project for life, there is nothing to do. Neither mothers nor the environment will provide the resources needed for survival under such extreme circumstances. Cultivating a posthuman wilderness ethic seems to be an alternative for the uncertain future where restorative justice will be granted to the human, nonhuman, and eco-systemic interactions.

In part II, Indigenous and Afro-diasporic myths and ecological knowledge, literature, anthropology, and popular culture contribute to a postcolonial and decolonial discussion of current situations of native and Afro-diasporic communities with special attention to their current aggravated living conditions. In a hemispheric movement that goes from the Canadian arctic zone—the land of the Inuit people—to a more temperate zone in Nova Scotia (Mik'maw), our journey traverses the United States, Puerto Rico and Haiti, and Brazil. Myth—as the bedrock of foundational ancestral stories that are communal and passed on generationally—usually engages with nature and the worlds beyond reality: spiritual, ritual, and symbolic. It crosses the limits between the human and the more-than-human. Animated by anticolonial sentiment and vindicating an autonomous ground to look into reality, the three major First Nations and ethnic populations represented in this section—Inuit, Mi'kmaqi, Puerto Rican and Haitian diasporas and Brazilians—use myth and ecological knowledges to preserve language, traditions, and ways of life in which the land, forest, and ocean build, maintain and re-shape communities.

In "Boundless water, boundless ice—arctic cosmological concepts in times of melting horizons," Sonja Ross addresses the impact of global warming on the lives of Inuit communities in Resolute Bay, on the Canadian Inuit territory of Nunavut. Within the Inuit, traditional ecological knowledges and mythological lore act as the foundation for the practices of everyday life. Climate change and its dramatic effects on communities from the north "defamiliarize" life as the Inuit people used to experience it. Driven by traditional cosmological views, star lore, and acute skills for observation, the Inuit have known how to move across space for survival, for subsistence hunting and resource use.

The Inuit people possess refined techniques of observation and therefore orientation. Moving through unknown terrain in the Arctic night requires special skills that only these peoples possess. However, the loss of semiotic–sensorial referents on the territory as a consequence of climate change damage the way of life of the Inuit.

By recourse to Inuit mythical narratives in which humans, animals, and other entities interact and co-exist, Ross shows how the "Inuit way of knowing" is idiosyncratic and particular, providing communities with relevant data for subsistence and continuity. Moreover, this chapter analyzes documentation on traditional ecological knowledges and questions widespread filmic interviews of a fair amount of Inuit people against the background of the aforementioned "Inuit way of knowing" that relies on concepts taken from mythological worldviews and also on the effects of environmental changes. Both of them feature as crucial pillars for survival.

In "Revisiting the wild: mythology and ecological wisdom in shalan joudry's *Waking Ground*," Leonor Martínez Serrano focuses on the Mi'kmaw natives from Nova Scotia. Ecological wisdom, a mythopoeic worldview, and the Mi'kmaw's attempts to counter the commodification of nature are part of their signs of identity. Poet shalan joudry, in her collection *Walking Ground*, places ecological concerns upfront. In her ecocentric worldview, she composes poems charged with mythological resonance to instill in readers a sense of ecological responsibility. Following eco-philosophical ideas, language becomes a property of animate earth, but it also serves to communicate with entities like rocks and trees.

Myth is a driving force that accounts for the origin and fascination with the world: cosmogonies and creation stories populated by characters, spirit beings, and form-shaping tricksters are distinctive in every community. Myth, like poetry and storytelling, is a form of knowing. They all play an active role in healing and recovery from the wounds of colonialism. In brief, shalan joudry uses language, myth, poetry, and resources like breathing and "walking ground" as "a true compass to human existence"—always understood in an almost symbiotic relation to the nonhuman—fusing the material and any other sphere for interaction as platforms for sustaining life.

In "Myth, Afrodiasporic spirituality, and the oceanic archive in independent comics," Paul Humphrey sets out to delineate how artists contributing to the cultural-cum-visual impulse within new comic book initiatives find nurture in the traditions of myth and spirituality of a wide Caribbean area—Puerto Rican and Haitian diasporas and Bahia, Brazil—that sustains life and opposes neocolonial practices such as extractivism and savage exploitation of resources and peoples.

This chapter makes for a stimulating reading of several independent comics, in which myth, spirituality, and the geographics of contact and exchange are shown in the web of the mythistorical oceanic archive. Incardinated within a Caribbean context where the memories of ancestral beliefs, forms of spirituality, and communal ritual practices still have a significant impact on life, these new visual art forms evince fundamental differences between these cultures and Western imperialistic imposed beliefs, rules, and laws. After showing how the Marvel universe has incorporated African and Afrodiasporic mythologies, the reader is introduced to knowledge of Candomblé, Vodou, Santeria, and Taino spiritualities. The coloniality of power/seeing is deconstructed by critically analyzing the potential that these spiritualities—strongly associated to the dialectic between the land and the ocean—show in their oppositional stances toward Western imperialism. Within the comics and graphic novel under discussion, the multispecies and embodied space of the ocean, badly damaged due to industrial and tourist exploitation, responds through the powers and marvelous deeds performed by its protagonists to counter Western colonialism and its appropriation of other cultures. In these, most times Caucasian characters are substituted by figures of West African and Afro-Brazilian mythology, where humans and more-than-human species interact. All through these comics and the popular culture associated to the persistence of ancestral memory and spiritual practices, the values of sustainability, reciprocity, and restoration are promoted and enhanced. These openings provide readers with ways of thinking, philosophies, and sensitivities that emerge from their own cultural archives.

In part III, Artistic practices, myth, and environmental resilience, the volume veers into art with a view on resilience, empowerment, and regeneration, through a journey that goes from Brazil to Morocco, Japan, and the US Southwest.

Along this culturally diverse scenario, myth bespeaks the pervasiveness of a cohesive narrative around an inexhaustible nature across time and space in Brazil, the passing on to the oral and material-cum-magical knowledge to communities in Morocco, and the power and intervention of myth into the regenerative drive that sustains life (and death) in Japan. With recourse to black studies and postcolonial critique and realities and notions such as the slave trade and its effects, the appropriation of land(scape), white exceptionalism versus savagery and barbarism, and ecocritical concepts such as environmental policies, recycling, and restoration and regeneration, this section demonstrates the critical power art has to raise awareness of the critical state of continued life on this planet.

In "'Giant by Thine Own Nature': Jean-Baptiste Débret and Antonio Parreiras's Mythic Brazilian Land(scape)s through a Transatlantic Gaze," Esther Lezra and Esther Sánchez-Pardo make an inquiry into the long-lasting myth of the Brazilian land and its pervasive presence in culture and social relations. They focus on the transition from Empire to Republic and on the evolution of Brazilian painting and its strong contribution to the representation of landscape and history. In the work of French painter and long-time resident in Brazil Jean-Baptiste Débret and of Brazilian painter Antônio Parreiras, a bountiful Brazilian land of mythical dimensions appears as an inexhaustible source of food, living space, commodities, shelter, and a distinctive national character. In a nuanced analysis of the evolution from the old times of the monarchy as captured by Débret in both his canvases and his book *Voyage Pittoresque et Historique au Brésil* and in the new Republic as portrayed in Parreiras's oeuvre, these two artists draw attention to racial and socio-ethnic relations when contact among indigenous populations, black slaves (brought through transatlantic trafficking), and white settlers becomes the norm. The clashes between social organizations, systems of values, and mythical worldviews show the gradual process through which the Brazilians come to inhabit an alternative modernity. From "virgin land" untouched by civilization—where myth, legend, and the enigmatic unknown occur—to the concrete sites where history takes place—as in Parreiras's "Conquista do Amazonas"—the countertropical operation inaugurates a moment for the construction of new myths for a new Brazilian nation at a distance from Western imposed models.

In "New cosmogonies of waste negotiated in the art of Mohamed Larbi Rahhali," María Porras Sánchez and Lhoussain Simour explore myth in connection with African everyday spirituality by addressing *baraka* in the artworks of Moroccan artist and fisherman Mohamed Larbi Rahhali. A pioneer in raising environmental awareness, Rahhali's works in the last 40 years are made of Mediterranean flotsam and discarded everyday objects, mostly matchboxes. The authors observe in Rahhali's installations a poetics of waste interlaced with mystic and mythic undertones, an environmentally aware cosmovision. As the authors show, Rahhali's practice of transforming waste into art becomes a symbolic recreation of *baraka*, "the mythic fabric of Moroccan spirituality and social life." Transforming a worthless item, such a matchbox, into an amulet that carries many cultural and environmental meanings is a form of artistic alchemy similar to the power of *baraka* (divine blessing) concentrated in charms or associated with certain sacred places and peoples. As an archive of memory, the environment, and the everyday, Rahhali's artworks become talismans that pass a blessing onto the viewers, creating an ecological and ethical engagement with them and proving the societal efficacy of art.

In "Death is life is death is life," Keijiro Suga elaborates on Japanese mythology and its close relation to nature with a focus on the work of a young painter and eco-artist, Maki Ohkojima (b. 1987). In Suga's view, contemporary art can be a philosophical praxis that invites audiences to rethink our living conditions. Japanese mythology, born of animism and the cult of every item in nature, is populated by many deities. It possesses an official lineage based upon stories woven to glorify the emperor and shows an outside perspective to this lineage, based upon stories invented and modified by communities with a keen sense for natural processes. The biological cycle of nature is a continuum in which life and death occur in quick succession. As myths feature repetition, they also become fixed and allow for our collective revision of the life-world. Upon explaining the autopoesis of the universe, the chapter focuses on Oogetushime as the goddess of food who, sacrificially killed, produces fecundity, because her dead body becomes a source of life. Drawing inspiration from this event, art and literature re-create stories and scenes in which new life emerges from death and decomposition.

In Ohkojima's art, the continuity of life and death is a crucial ground for her creative process. Her work re-creates the biological cycle and elaborates on metamorphoses, scenes of transformation, interchangeabilities (eg., forest and body), and correspondences—referring to Baudelaire's synesthesia, as well as to connectivities at large. From Ohkojima's art and her injunction, "we need to imagine the entire earth as if it were our own body," the evidence of the continuity between myth and environmentalism remains clear.

Finally, the Coda to our volume, "A radical evocation of seed," is an artistic statement by artist, farmer, and scholar Jeanette Hart-Mann in which she explores her ongoing eco-social praxis, started in 2007 in the American Southwest, in which she works with seeds. Hers is a transdisciplinary example of "sympoiesis" in which her commitment with art, agriculture, activism, and environmental and social justice coalesce in the recollection of seed stories. In her different artworks—installations including photographs, videos, and acoustic recordings—seeds are more-than-human beings that challenge the mythos of capitalism and colonialism. Their radical evocation is, in Hart-Mann's words "a call to action … to listen deeply, bear witness, and make meaning together in order to respond to our collective ecological crisis." As an ecological imaginary, corn in particular is considered a "mythological mother" that recalls the struggles and suffering of indigenous communities, the dangers of genetically modified organisms, and also family and community. Describing her latest artworks, Hart-Mann opts for a decolonial approach to heal trauma created by settler-colonial past by resituating the seed story of corn and preserving and growing Ohio Corn Belt dents.

Her appreciation of the value of the land—especially when dealing with the Southwest—is narrowly linked to Amerindian cosmovisions. Hart-Mann's Coda is an important performative enactment of the ways in which innovative and sustainable art practices are intertwined with myth and environmentalism.

Throughout this journey, our volume explores new ground at the interface between the more classical world of myth and the urgent present of environmental decay. In a global perspective, the historical and cultural specificity of mythical narratives provides instances of profound entanglements with nature at large. From stories mostly passed down from oral traditions to other formats and media such as poetry, video art, painting, comics, novels, and documentaries, a much-needed transdisciplinary cooperation emerges in the struggle against the depletion of resources and destruction of the planet. *Myth and Environmentalism* aims to contribute to this ongoing endeavor.

Notes

1 Environmentality is a term used to describe an approach that builds on Michel Foucault's concept of governmentality developed in the early 1980s (Foucault 1991). According to Arun Agrawal, environmentality is "a framework of understanding in which technologies of self and power are involved in the creation of new subjects concerned about the environment'" (2005, 166).
2 David Abram coined the term "more-than-human." He claims, "How is it we have become so deaf and so blind to the vital existence of other species, to the animate landscapes they inhabit that we now bring about their destruction?" (1996, 28–29).
3 According to Odum and Barrett, "Haeckel defined ecology in 1869 as the study of the natural environment including the relations of organisms to one another and to their surroundings" (2004, 3).
4 See, for instance, the COP26 Glasgow Summit Pact (UNFCC 2021).
5 Masao Miyoshi (2001) and Wai Chee Dimock (2006) have been major voices in the debate on planetarity. See also Elias and Moraru (2015) for an interesting update on different positions on the planetary turn.
6 With "deep time," James McPhee (1981) referred to geological time, and thus, to the research that involves the age of Earth, approximately 4.5 billion years.

7 By "recognition gap," Lamont argues for "[…] promoting a plurality of criteria of worth and 'ordinary universalism,' as well as the destigmatizatiton of stigmatised groups" (2019, 667).
8 We owe to anthropologist Gananath Obeyesekere the crucial notion "work of culture." In his homonymous book (1990), Obeyesekere explores what he calls "symbolic remove"—"the process through which symbolic forms existing at the cultural level. are created and recreated through the minds of individuals. Symbols thus created are regressive because of their ontogenesis in individual development and unconscious processes, while also being progressive, in that the unconscious thought transforms the archaic motivations of early experience and looks forward to their realization in the experience of the sacred" (Nuckolls 1997, 133).
9 Hayden White worked extensively on emplotment from his crucial *Metahistory* (1973) onwards. In his view, to the four tropes mostly responsible for imbuing past events, metaphor, antonymy, synecdoche, and irony, there are four corresponding generic structures or modes of emplotment: romance, tragedy, comedy, and satire.
10 This has been discussed by Indian activist and eco-feminist Vandana Shiva in *Monocultures of the Mind* (1993).
11 In his important essay "On the Museum's Ruins" (1980; published as a homonymous book in 1993), Douglas Crimp, in a Foucaultian vein, argued against museums remaining institutions of confinement similar to asylums and prisons in the French philosopher's crucial analysis.

References

Abram, David. 1996. *The Spell of the Sensuous: Perception and Language in a More-than-Human World.* New York: Vintage.

Agrawal, Arun. 2005. "Environmentality: Community, Intimate Government, and the Making of Environmental Subjects in Kumaon, India." *Current Anthropology* 46, no. 2: 161–190.

Alexander, Jeffrey C. 2004. "Cultural Pragmatics: Social Performance between Ritual and Strategy." *Sociological Theory* 22, no. 4: 527–573.

Allen, Paula Gunn. 2015. *The Sacred Hoop: Recovering the Feminine in American Indian Traditions.* New York: Open Road (e-book).

Athens, Allison K. 2013. "*Arctic Ecologies: The Politics and Poetics of Northern Literary Environments.*" Ph.D. thesis, University of California–Santa Cruz.

Barber, Elizabeth, and Paul Barber. 2004. *When They Severed Earth from Sky. How the Human Mind Shapes Myth.* Princeton, NJ: Princeton University Press.

Barthes, Roland. (1957) 1972. *Mythologies*, translated by Annette Lavers. London: Paladin.

Barthes, Roland. 1997. "Bichon among the Blacks." In *The Eiffel Tower and Other Mythologies*, translated by Richard Howard, 35–38. Berkeley: University California Press.

Bauman, Zygmunt. 2001. *The Individualized Society.* Cambridge: Polity.

Bauman, Zygmunt, and Leonidas Donskis. 2015. *Moral Blindness: The Loss of Sensitivity in Liquid Modernity.* Cambridge: Polity.

Bauman, Zygmunt, and Ricardo de Querol. 2016. "Interview." *El País. Babelia*, January 21.https://elpais.com/cultura/2015/12/30/babelia/1451504427_675885.html.

Bruni, Gilles, and Marc Babarit. 2004. "Eco-Habitat." In *Art Nature Dialogues. Interviews with Environmental Artists*, edited by John K.Grande, 25–32. Albany: SUNY Press.

Caminero-Santangelo, Byron, and Garth Myers, eds. 2011. *Environment at the Margins: Literary and Environmental Studies in Africa*. Athens: Ohio University Press.
Carson, Rachel. 1962. *Silent* Spring. Boston: Houghton Mifflin.
Chakrabarty, Dipesh. 2009. "The Climate of History: Four Theses." *Critical Inquiry* 35, no. 2: 197–222.
Chakrabarty, Dipesh. 2021. *The Climate of History in a Planetary Age*. Chicago: University of Chicago Press.
Chandler, David, Kevin Grove, and Stephanie Wakefield, eds. 2020. *Resilience in the Anthropocene: Governance and Politics at the End of the World*. London: Routledge.
Chepesiuk, Ron. 2007. "Environmental Literacy: Knowledge for a Healthier Public." *Environmental Health Perspectives* 115, no. 10: A494–A499. https://www.ncbi.nlm.nih.gov/pmc/articles/PMC2022675/.
Code, Lorraine. 2006. *Ecological Thinking: The Politics of Epistemic Location*. Oxford: Oxford University Press.
Crews, Chris. 2019. "Earthbound Social Movements and the Anthropocene." *Journal for the Study of Religion, Nature and Culture* 13, no. 3: 333–372.
Crimp, Douglas. 1993. *On the Museum's Ruins*. Cambridge, MA: MIT Press.
DeLoughrey, Elizabeth, and George Handley. 2011. *Postcolonial Ecologies: Literatures of the Environment*. New York: Oxford University Press.
Dimock, Wai Chee. 2006. *Through Other Continents. American Literature across Deep Time*. Princeton, NJ: Princeton University Press.
Dougherty, Patrick. 2004. "Yardworking." In *Art Nature Dialogues. Interviews with Environmental Artists*, edited by John K. Grande, 15–24. Albany: SUNY Press.
Elias, Amy, and Christian Moraru, eds. 2015. *The Planetary Turn. Relationality and Geoaesthetics in the 21st Century*. Evanston, IL: Northwestern University Press.
Evernden, Neil. 1992. *The Social Creation of Nature*. Baltimore, MD: Johns Hopkins University Press.
Foucault, Michel. 1991. "Governmentality." In *The Foucault Effect: Studies in Governmentality*, edited by Graham Burchell, Colin Gordon, and Peter Miller, 87–104. Chicago: University of Chicago Press.
Fowler, Robert L. 2011. "Mythos and Logos." *The Journal of Hellenic Studies* 131: 45–66.
Friedman, Susan Stanford. 2018. *Planetary Modernisms: Provocations on Modernity across Time*. New York: Columbia University Press.
Geertz, Clifford. 1973. *The Interpretation of Cultures. Selected Essays*. New York: Basic Books.
Haraway, Donna. 1988. "Situated Knowledges: The Science Question in Feminism and the Privilege of Partial Perspective." *Feminist Studies* 14, no. 3: 575–599.
Haraway, Donna. 2014. "Anthropocene, Capitalocene, Chthulucene: Staying with the Trouble." http://opentranscripts.org/transcript/anthropocene-capitalocene-chthulucene/.
Highway, Tomson. 2003. *Comparing Mythologies*. Ottawa: University of Ottawa Press.
Highway, Tomson. 2022. "Aboriginal Mythology." https://tomsonhighway.com/speaking-performing/.
Huggan, Graham, and Helen Tiffin. 2010. *Postcolonial Ecocriticism: Literature, Animals, Environment*. London: Routledge.
Hull, Lynne. 2007. "Eco-Art eConversations." https://ecology-artelecture.blogspot.com/2007/04/elementary-lesson-plan-nature-is-art.html?m=0.
IPCC. 2013. *Climate Change 2013: The Physical Science Basis. Working Group I Contribution to the Fifth Assessment Report of the Intergovernmental Panel on Climate Change*. Cambridge: Cambridge University Press.

Jandl, Nathan. 2013. "*Counter-Love: The Social Dimensions of Environmental Attachment in Twentieth-Century American Literature.*" Ph.D. thesis, University of Wisconsin–Madison.

Kolbert, Elizabeth. 2014. *The Sixth Extinction: An Unnatural History.* New York: Henry Holt.

Kolodny, Anette. 1975. *The Lay of the Land. Metaphor as Experience and History in American Life and Letters.* Chapel Hill: University of North Carolina Press.

Kuipers, Giselinde. 2019. "Cultural Narratives and Their Social Supports, or: Sociology as a Team Sport." *British Journal of Sociology* 70, no. 3: 708–720.

Lamont, Michele. 2019. "From 'Having' to 'Being': Self-Worth and the Current Crisis of American Society." *British Journal of Sociology* 70: 660–707.

Latour, Bruno. 2004. "Whose Cosmos, Which Cosmopolitics?: Comments on the Peace Terms of Ulrich Beck." *Common Knowledge* 10, no. 3: 450–462.

Martinez-Alier, Joan, and Lorenzo Pellegrini. 2012. "Interview." *Development and Change* 43, no. 1: 341–359.

Marx, Leo. 1964. *The Machine in the Garden: Technology and the Pastoral Ideal in America.* Oxford: Oxford University Press.

McBride, B.B., C.A. Brewer, A.R. Berkowitz, and W.T. Borrie. 2013. "Environmental Literacy, Ecological Literacy, Ecoliteracy: What Do We Mean and How Did We Get Here?" *Ecosphere* 4, no. 5: 1–20. https://doi.org/10.1890/ES13-00075.1.

McPhee, James. 1981. *Basin and Range.* New York: Farrar, Straus and Giroux.

Merriam-Webster. 2022. "Environment." *Merriam-Webster Dictionary.* https://www.merriam-webster.com/dictionary/environment.

Miyoshi, Masao. 2001. "Turn to the Planet. Literature, Diversity and Totality." *Comparative Literature* 53, no. 4: 283–297.

Naredo, José Manuel. 2006. *Raíces económicas del deterioro ecológico y social: más allá de los dogmas.* Madrid: Siglo XXI.

Nixon, Rob. 2005. "'Environmentalism and Postcolonialism." In *Postcolonial Studies and Beyond*, edited by Ania Loomba, Suvir Kaul, Matti Bunzl, Antoinette Burton, and Jed Esty, 233–251. Durham, NC: Duke University Press.

Nuckolls, Charles W. 1997. "Review of the Book *The Work of Culture: Symbolic Transformation in Psychoanalysis and Anthropology*, by G. Obeyesekere." *Psychoanalytic Psychology* 14, no. 1: 133–134.

Obeyesekere, Gananath. 1990. *The Work of Culture: Symbolic Transformations in Psychoanalysis and Anthropology.* Chicago: University of Chicago Press.

Odum, Eugene, and Gary Barrett. 2004. *Fundamentals of Ecology.* Boston: Cengage Learning.

O.E.D. 2022, Environment, n." *OED Online.* www.oed.com/view/Entry/63089.

Oring, Elliott. 1986. "Folk Narratives." In *Folk Groups and Folklore Genres: An Introduction*, edited by Elliott Oring, 121–146. Logan: Utah State University Press.

Pérez Orozco, Amaia. 2014. *Subversión Feminista de la Economía. Sobre el Conflicto Capital-Vida.* Madrid: Traficantes de Sueños.

Plumwood, Val. 1993. *Feminism and the Mastery of Nature.* London: Routledge.

Roos, Bonnie, and Alex Hunt, eds. 2011. *Postcolonial Green: Environmental Politics and World Narratives.* Charlottesville: University of Virginia Press.

Said, Edward. 1993. *Culture and Imperialism.* New York: Vintage.

Schiff, Sarah Eden. 2010. "*Word of Myth: Critical Stories in Minority American Literature.*" Ph.D. thesis, Emory University.

Shiva, Vandana. 1993. *Monocultures of the Mind: Perspectives on Biodiversity and Biotechnology.* London: Zed Books.

Siegert, Martin J. 2016. "Environmental Sciences in the Twenty-First Century." *Frontiers in Environmental Science* 4: 16. doi:10.3389/fenvs.2016.00016.
Slotkin, Richard. 1973. *Regeneration through Violence: The Mythology of the American Frontier, 1600–1860*. Middletown, CT: Wesleyan University Press.
Slotkin, Richard. 1992. *Gunfighter Nation: The Myth of the Frontier in Twentieth-Century America*. New York: Atheneum.
Spivak, Gayatri Chakravorty. 2003. *Death of a Discipline*. New York: Columbia University Press.
UNFCC. 2021. *COP26: The Glasgow Climate Pact*. https://ukcop26.org/wp-content/uploads/2021/11/COP26-Presidency-Outcomes-The-Climate-Pact.pdf.
Wagner-Lawlor, Jennifer A. 2017. "Plastic/ity." In *Symptoms of the Planetary Condition. A Critical Vocabulary*, edited by Mercedes Bunz, Birgit Mara Kaiser, and Kathrin Thiele, 97–101. Lüneburg, Germany: Meson Press. https://doi.org/10.25969/mediarep/2011.
Weintraub, Linda. 2012. *To Life: Eco Art in Pursuit of a Sustainable Planet*. Berkeley: University of California Press.
White, Hayden. 1973. *Metahistory: The Historical Imagination in Nineteenth-Century Europe*. Baltimore, MD: Johns Hopkins University Press.
Zelnik, Eran. 2016. "*The Comical Style in America: Humor, Settler Colonialism, and the Making of a White Man's Democracy, 1740–1840*." Ph.D. thesis, University of California–Davis.

PART I

Myth, disaster, and present-day views on ecological damage

1

THE AFTERLIFE OF CHORNOBYL

Apocalyptic mythology and environmentalism in the Exclusion Zone

Haley Laurila

When an explosion shattered the containment structure for reactor 4 of the Chornobyl Nuclear Power Plant in Ukraine in late April of 1986, the subsequent catastrophe was apocalyptic.[1] Many Ukrainians saw the disaster as divine punishment for their complacency and passivity in allowing the disaster to occur. The magnitude of the disaster necessitated the creation of a 39-km Zone of Exclusion (*Zona Vidchuzheniya*), which is cordoned off from casual entry and habitation. Ukrainian stalkers explore the Zone illegally, challenging the ideas of death, decay, and ruin that are typically associated with the abandoned space. They participate in its demythologization by showing both the stark ecological and psychological realities of nuclear disaster but also the potential for hope and change within revelation. In charting this postapocalyptic terrain, stalkers amass intimate knowledge of the Zone's nuclear landscape, and their engagement with Chornobyl's apocalyptic mythology reveals pertinent instructions for our own precarious future.[2] This chapter will examine the varied mythological dimensions that frame this disaster through the contrasting but resonant experiences of two Ukrainians, writer Markiyan Kamysh and artist Fedor Oleksandryvych, who both claim a strong relationship with the Zone. Not intended to serve as an analysis of their works, this chapter instead explores how their experiences offer insight into a future created by our egregious exploitation of the environment. They reveal how apocalyptic thinking can help us recognize Chornobyl's long-lasting impacts on the environment as well as the specific historical contexts and ideologies that sustain ecological harm. Apocalyptic ruptures reveal opportunities for us to unapologetically commit to the care and protection of the Earth within a more expansive and ecologically informed network of responsibility for both human and nonhuman others.

DOI: 10.4324/9781003348535-3

Soviet mythologies: the veneration of nuclear power

The catastrophic explosion at the Chornobyl Nuclear Power Plant released 200 million curies of radioactive material into the air and onto the surrounding landscape. Although Soviet authorities attempted to contain the disaster with assurances that the situation was under control and by limiting the scope of information divulged to the public, it quickly became clear that total containment was impossible. Hundreds of thousands of Soviet workers were mobilized to douse the dangerous fires, which burned for days, and to clean up the radioactive debris. Inhabitants of the city of Prypiat and its surrounding villages were evacuated, their homes were leveled, their pets and livestock killed, and their gardens and crops buried. Families that had worked the land for generations found themselves "torn from the land and graves of their ancestors" (Stepanets 2017, 98, translation mine). The official number of deaths caused by Chornobyl—just 31—does not account for the deaths from radiation exposure. Some estimates put that number into the hundreds of thousands, but the final count of victims may never be known. Despite the lack of long-term health monitoring, the preliminary data collected in the immediate aftermath of the disaster indicate a growing public health crisis given the long half-life of some radioactive isotopes and the elevated rates of cancer associated with the exposure (Brown 2019). Not only did Chornobyl leave its radioactive trace on both bodies and spaces but it also led to the collapse of the Soviet Union and cast suspicion on any visions of a global nuclear-powered future.

Before dissecting Chornobyl's subsequent mythologization as a postapocalyptic wasteland, it is important to understand how the event challenged dominant Soviet cultural mythologies and laid bare the infrastructures of power responsible for the exploitation of land across the Soviet Union. These mythologies projected an impossible techno-utopian communist future that could never exist. At the core of the development of nuclear power is the myth of the "peaceful atom" (*mirnyi atom*). The peaceful atom is not a Soviet concept but one employed around the globe to promote the safety of nuclear power and project utopian fantasies (Josephson 1996). From the time of Lenin, these utopian fantasies fueled the construction of Soviet society and transformed agricultural Russia into a modern, industrialized Soviet nation. It is a familiar story, as Andy Bruno notes in his study of environmental entanglements in the Soviet north:

> Around the globe, governments of the last century usually privileged unrelenting economic expansion over limiting the pressures placed on natural systems. If anything, the Soviet Union followed this modern growth imperative more eagerly than most, due, at least initially, to an acute sense that it needed to rapidly overcome the country's supposed backwardness.
>
> *(2016, 16)*

This aggressive drive for progress and technological advancement was channeled into ambitious production plans and huge industrial projects that often involved subverting the course of nature. In their discussion of Soviet ecocide, Feshbach and Friendly conclude: "Ecocide in the USSR stems from the force, not the failure, of utopian ambitions" (1992, 29). The lack of an outspoken civic culture meant that opposition to environmentally harmful enterprises within the Soviet Union was often silenced (Josephson et al. 2013). Huge hydroelectric power stations, massive mining operations, oil and resource extraction, and irrigation projects demonstrated the prowess of Soviet science and engineering, but mastery of the atom was to be its crowning achievement. Scientists, eager to explore the possibilities of this new technology, lobbied the government for more investment in nuclear energy, presenting it as a viable solution to future energy demands, a sign of Soviet superiority to the capitalist West, and a means of modernizing remote regions of the country (Schmid 2015).

The myth of the peaceful atom in the Soviet Union, as historian Paul Josephson explains, was borne out of a robust cult of science and technology bolstered by an "unbounded faith in science and technology" (1996, 299). Nuclear culture was represented in a visual culture promoting the atom's contribution to "the betterment of humankind" and as "essential for national economic and social programs" (298). Nuclear physicists and engineers occupied an elite position within the Soviet bureaucracy and enjoyed the high status afforded by their important role in the construction of communism. Though not an exclusively Soviet condition, Soviet nuclear science operated under a veil of secrecy with minimal external oversight, which lent the entire nuclear enterprise an unwarranted arrogance. Soviet nuclear culture transformed a complex, nuanced science into a standardized process that could be quickly exported to the republics to achieve the impossible political and economic expectations of modernizing such a vast territory. From its start in 1964, the process of civilian nuclear development released huge amounts of radiation into the environment and onto unsuspecting citizens, so that contamination became "quotidian across the USSR" (Brown 2019, 136). Unsurprisingly, before the true costs of the industry's development were exposed, nuclear power had become a national priority, and its spread throughout the Soviet republics was hailed as a triumph of socialism. Given the mythological dimensions of nuclear futurity, it is no wonder that religious imagery, as Elena Romashko explains, "appealed to long-playing utopian and apocalyptic allusions and became incessant leitmotifs in artistic perceptions of Chernobyl" (2016).

In Ukraine, local authorities fully embraced the promises of modernization offered by the nuclear industry. Ukraine was a "comfortable home" for these kinds of efforts, particularly given the republic's "physical proximity to Russia" and "strong tradition of nuclear physics" (Josephson 2005, 204). The blame that many Ukrainians felt after Chornobyl stems from recollecting the enthusiasm displayed by Ukrainian party leaders in welcoming nuclear power and assuaging fears with propaganda. Ukrainian poets, some who were pressured by the threat of censure and silence, wrote plays, poems, and novels celebrating the coming of nuclear power and the radiant communist future. Others genuinely believed in the opportunities and benefits promised by nuclear progress. Ivan Drach, a prominent poet with a complicated relationship with the state, wrote a series of poems venerating the Chornobyl NPP and nuclear power in his 1974 collection. One poem, "The Legend of Polissia" ("Poliiska Lehenda"), tells of the marriage between the river-bride Nature and her suitor Atom. Nature must convince her people that their union will bring prosperity to the Polissia region and communities on the Dnipro River. Drach draws on familiar elements from folklore and nature to neutralize the atom's destructive associations for Ukrainian readers. These same Polissian communities, who were dependent on this land for farming, suffered the most from the nuclear disaster: "Farmers didn't invent Chernobyl, they had their own relations with nature, trusting relations, not predatory ones, just like they had one hundred years ago, and one thousand years ago" (Alexievich 2006, 173). Vasyl Shevchuk, Ukraine's former minister of the environment and deputy minister for sustainable development, similarly explains that this region "has a special spirituality, a special national value" with strong and deep cultural roots: "The Dnipro and Ukraine—these are the father and the mother of the Ukrainian nation" (Shevchuk et al. 2005, 8). Ukrainian stalker Kyrylo Stepanets notes that the region had previously been subject to ecological harm when the course of the Uzh river was diverted with heavy explosions, disrupting the local ecosystem. Residents were already inclined to "believe that the land was cursed from the beginning," even then recalling the apocalyptic mythology of St. John in reference to previous Soviet incursions (Stepanets 2017, 98, translation mine). In further confirmation of the region's significance, well-known Ukrainian poet Lina Kostenko led expeditions into the villages of the Chornobyl Exclusion Zone to preserve the region's cultural heritage, calling it the "Polissian Atlantis" (Mykoliuk 2009). For Kostenko, who was born in the region, the technogenic origins of the disaster are secondary to the cultural history contained there.

Chornobyl was not an end but the beginning of an independent Ukrainian state. Fueled by glasnost, the damage accrued by the nuclear disaster ignited a politically charged environmental movement against "Moscow's eco-imperialism" (Plokhy 2018, 305). Environmental justice and national sovereignty were linked because the disaster was seen as justification for Ukrainian statehood. In his history of the disaster, historian Serhii Plokhy details Ukrainian writers' and intellectuals' efforts to turn environmental fears into meaningful action. Some were partly motivated by the need to atone for their own roles in promoting nuclear power. Writers such as Ivan Drach and Volodymyr Yavorivsky, who had initially celebrated the coming of nuclear power as a sign of modernity for Ukraine in their novels and poetry, quickly turned into vocal political activists. Poet Borys Oliinyk offered scathing remarks on the environmental exploitation of Ukraine at a conference of the Writers Union in 1988, calling out the "arrogance and neglect of several union bodies, and above all the Ministry of Energy, to the fate of Ukraine borders on merciless cruelty and with an insult to national dignity" (Oliinyk 1988, translation mine). He continues, reminding the audience that their valid concerns and objections to the plant's construction had been met with disdain and ridicule. These writers became environmental activists, demanding political accountability, the closure of all nuclear power plants in Ukraine, and increased health monitoring for affected regions. In the words of Drach, the apocalyptic nature of Chornobyl had "roused our souls, showing us in real terms that we were on the edge of a precipice, an abyss" (cited in Plokhy 2018, 293). The apocalyptic nature of Chornobyl revealed the failings and problems of Soviet institutions and ideology. As traumatic as the disaster was and continues to be, it reinvigorated a long-held desire to throw off the yoke of Soviet nuclear colonialism.

Apocalyptic thinking and the importance of revelation

For many Ukrainians, the Chornobyl catastrophe was apocalyptic and interpreted as divine punishment for the hubris of mankind's nuclear ambitions and the resulting ecological harm inflicted on "God's domain" of nature (Phillips 2008, 163). The catastrophe recalled the prophecy told by St. John the Divine in the Book of Revelation, which reads:

> Then the third angel blew his trumpet, and a large star, burning like a torch, fell from the sky. … The name of the star is Wormwood. And a third of all the water became bitter, and many people died from drinking the water that was bitter.
>
> *(New International Version 2011, Revelation 8:10–11)*

Wormwood is the third sign of the apocalypse, "a frequent symbol for bitterness, calamity, and sorrow" (Mycio 2005, 8). The connection to this biblical prophecy hinges upon the translation of *chornobyl'* in Ukrainian to "wormwood." A more accurate translation for the *chornobyl'* plant is mugwort, which is a relative of the actual wormwood plant, or *polyn gorkaia*. Proper botanical nomenclature aside, the accuracy of translation does not detract from the power of the association, nor the imagery and cultural myths that sprouted and took root. Journalist Mary Mycio notes that the apocalyptic fears associated with the disaster became so prevalent that interviews with Russian Orthodox Church clergy were aired to reassure the public (2005). Still, the association remained because, as Elena Romashko contends, the religious association "is still deeply integrated into nuclear culture" and can, in turn, help us respond to and challenge the primacy of nuclear power (2019, 193–194).

The prevalence of this interpretation can be further detected in Chornobyl poetry, art, and literature, as well as in Chornobyl's memorials and museums. In the Wormwood Star Memorial Complex in the town of Chornobyl sits Anatolii Haidamak's "Monument of the Third Angel," a direct reference to the biblical apocalyptic prophecy. This memorial sculpture sits amidst other memorials that lament the disaster and remember the victims. Volodymyr Yavorivsky's novel *Maria with the Bitter Wormwood at the End of the Century* (1987) uses the prophecy as a framework for the story of Maria, a mother who loses two of her three sons to the disaster. Larissa M.L. Zaleska Onyshkevych, who analyzed several literary works on Chornobyl, discerns "a guilt syndrome" in early post-Chornobyl literature in which "Chornobyl takes the form of moral punishment for the community, albeit at the cost of many innocent lives" (1990, 284). Anatoliy Mykhailenko's novel *The Smell of Wormwood* similarly invokes the biblical prophecy and presents the Ukrainian people as "historically guilty" (1990, 286). Literary critic Tamara Hundorova expands on this idea further, linking Chornobyl's apocalyptic reverberations to the emergence of a Ukrainian postmodernism in which "the post apocalyptic situation is described" (2019, 22). This act of description encompasses how writers document the fallout from the disaster's shattering of a monolithic reality into a multiplicity of ambiguous and shifting realities. The "Chornobyl text," as Hundorova terms the disaster's discourse, is characterized by apocalyptic thinking and is marked by an explicit suspicion of "universal progress and the development of science, reason, and technology, as well as the idea of socialism itself" (21). Amidst the carnivalistic upheaval caused by nuclear disaster, the opportunity to reinvent and remake reality emerged.

Andrea Oppo, who has written widely on Russian philosophy, boldly claims that the myth of apocalypse is a Slavic preoccupation characterized by "a specific contemplation of the concepts of 'end of present time' and 'end of history' as conditions for a redemptive image of the world" (2013, 9). The redemptive element originates in the conception of time *kairos*, which refers to an in-between time, or "the opportune moment, the supreme time to do something" (Oppo 2013, 23). In contrast, Western iterations of apocalypse tend to rely on a temporal organization based on *chronos*, or the linear chain of events and history as we know it, which imagines a final ending. The perspective of *kairos* embodies the cyclical potential in apocalypse that lays bare the present time to bring about transformation (Oppo 2013). Johanna Lindbladh shows how apocalyptic narratives in Chornobyl films tend to emphasize the potential for rebirth and revelation, which also becomes "important also on a collective level, namely in terms of a mythical discourse, emphasizing a cyclic, rather than a chronological, perception of history" (2019, 241). Examining Chornobyl films from Ukraine, Belarus, and Russia, Lindbladh finds that the Chornobyl disaster, even in its trauma, loss, and destruction, is represented as a "positive" force "in the sense that this apocalyptic event contributes to a moral, religious, political, existential or emotional awakening" (253). From this perspective, the villages and towns that were evacuated in the Zone of Exclusion become "nostalgic, mythical, lost homes, to which the characters want to return" even as they are forced to abandon them (Lindbladh 2019, 254). In reality, Chornobyl fractured any security and trust that had been promised by the Soviet myth of a bright and prosperous future and victims "describe the experience of Chernobyl fallout and its destructive closeness as radically alienating them from a sense of security in public space, in their home and in their own bodies."(Barcz 2021, 135). The average Soviet citizen had neither the expertise nor the equipment to adequately measure the extent to which radiation had penetrated their bodies and environment. In these confusing and uncertain times, it is unsurprising that average citizens turned to the apocalyptic, which "thrives in a climate of ambiguity of subtext, innuendo, and insinuation" (Riedl and Marno 2018, xii). Additionally, as Romashko explains, the immense energy associated with nuclear power "makes its religious and mythological connotations unavoidable" (2016). Amidst the senselessness of nuclear disaster, apocalyptic mythology's revelatory power opens a space for reevaluation and action that "gives hope even in unjust and destroyed worlds" (Romashko 2016).

Lilya Berezhnaya's analysis of Ukrainian millennialism adds necessary nuance to Oppo's assertions as to the Slavic preoccupation with apocalypticism. She cautions against grouping Ukrainian culture under the umbrella of Russian thought, claiming that "an eschatological frame of mind is by no means a national attribute" among Ukrainians but, rather, this kind of thinking is "mostly fueled by uncertainty, despair, disorder, and the breakdown of the traditional social order" (Berezhnaya 2018, 272). Though she does acknowledge the difficulties in isolating a distinctly Ukrainian eschatological history given the proximity of the two countries, Berezhnaya maintains that much of Ukraine's apocalyptic resurgences are connected to the struggle to gain independence from either Polish, Russian, or Soviet control. Still, the nature of apocalypse in the Ukrainian context must embody the idea of renewal out of necessity, given the historical traumas it has endured in the struggle for statehood. The 1991 push for independence was informed by the realization that the ecological trauma caused by exploitative Soviet industrial practices and epitomized by Chornobyl, could no longer be endured. In contrast, the enduring nature of radiation makes the environmental damage of nuclear disasters particularly difficult to recover from. With a half-life of thousands of years, some of the radioactive material housed in the decommissioned Chornobyl NPP, as well as radiation in the water and soil, will remain dangerous for the foreseeable future. Unfortunately, Chornobyl is also part of a larger, more devastating Soviet nuclear legacy that includes years of radioactive waste dumping in the Techa River, a radioactive explosion in 1957, and a radioactive dust storm in 1967, all traced to operations at the Mayak nuclear facility in Chelyabinsk, Russia (Whiteley 1999). Add to this the Soviet Union's atomic bomb testing in Kazakhstan and it becomes clear that the history of the Soviet Union is a history of ecological fracture and crisis rife with apocalyptic events. These nuclear catastrophes show how nuclear power's "only disposition is to produce futurelessness for beings, ecosystems, and planets" (Hurley 2020, 210). The Chornobyl nuclear disaster itself may not have caused mankind's end, but its lasting consequences alert us to the possibility of a world without humans.

The environmental exploitation that accompanied Soviet progress is not unfamiliar. Chornobyl and the entire Soviet nuclear legacy are part of a global constellation of violence inflicted on bodies and spaces across the planet. Forever coupled with atomic bomb manufacturing, the history of nuclear power is one of extensive environmental and epidemiological harm. Jessica Hurley, in discussing the American nuclear complex through the writings of neglected authors, incriminates the global nuclear infrastructure for promoting a dangerous investment in nuclear power as either a source of real political power or a climate change panacea. She asserts that nuclear apocalypse "requires a planetary infrastructure to bring it about and a deep and ongoing commitment to maintaining its possibility" (2020, 13). Chornobyl is one just one node in a web of nuclear violence in which unsuspecting civilians are acutely impacted by its misuse as others profit from nuclear development. Nuclear contamination is what Rob Nixon terms a "slow violence" and is part of mankind's legacy in this Anthropocenic epoch of ecological fracture and crisis (2011, 260). Whether global warming, air and chemical pollution, toxic microplastics, deforestation, or nuclear waste, the accumulated damage we have wrought on the environment in the name of progress has left us with a suffering planet. The story of Chornobyl is also a metanarrative of the larger ecological crises in which we find ourselves. Apocalyptic thinking, in its emphasis on revelation, can facilitate the necessary envisioning of an alternative eco-conscious future if only we choose to bear witness to the cycle of destruction we have already enacted.

Stalker culture and the demythologization of the past

It is important to remember that Chornobyl is not only an apocalyptic event but also a real place. No longer just the vernacular name for the nuclear power plant or the name of a village outside of Prypiat, "Chornobyl" is also the 30-km Zone of Exclusion. However, even the boundaries of the Zone are somewhat mythical given that contaminated areas of Belarus are not included but are still inhabited.[3] Within the Zone, modern ruins offer material evidence of the fallibility of nuclear techno-utopian dreams. A new confinement structure now covers the old sarcophagus that was built over the exploded reactor. There are villages, some completely buried and lost to us, while others, aided by looters, slowly crumble. Artifacts of Soviet culture are scattered throughout empty, collapsed buildings obscured by weeds. The broad boulevards of the planned atomic city of Prypiat are overgrown; wild animals prowl the streets and make their home there. Radioactive hotspots dot the Zone, as well as radioactive dust. No one is allowed to live in the Zone, although a small contingent of illegal settlers have returned to their contaminated homes and land. Workers monitor and maintain the defunct power plant and aid in its decommissioning, and police patrol the Zone for looters and stalkers. The Zone has also become a tourist destination: official tours are the only legitimate means of visitation, and thousands of tourists visit the Zone each year. Depictions of the Zone as a deserted wasteland do not offer an adequate account of how this space functions today, nor do they fully account for the kinds of transformations taking place there.

In the decades since the reactor's explosion, the Zone has experienced a resurgence of nature. The once-neat gardens and cultivated fields have run wild. New trees and vegetal growth fill the expanding cracks in the concrete and weave around rusted metal structures. Wild boars, elk, dogs, wolves, foxes, and bears roam the Zone, and there is even a growing herd of Przewalski's horses, which were introduced in 1998. To the casual observer, it would be too easy to say that the Zone has recovered and the danger is gone, having been replaced by a nature preserve, as in the case of the Palieski State Radioecological Reserve, which makes up a large portion of the Belarusian side of the Exclusion Zone. As Michael Marder writes, "Plants will gently gag the silent scream of things. Where there was devastation and abandon, there will be a forest" (2016, 58). Ironically, the trees and plants help make the region safer by absorbing the ionizing radiation, even as it damages and mutates the plant's genetic structures. The Zone's tentative recovery is connected more so to the absence of humans living there than the absence of radiation, which is telling. Though spending time in the Zone on an official tour is not dangerous, there are areas that should be avoided due to the high levels of radiation still present. Certain objects absorb radiation better than others, and some plants, like mushrooms, retain radiation longer. The infamous Red Forest (*Chervonyi Lis*) received a large amount of radiation that turned the pine trees red as they died. Much of the forest was destroyed and buried long ago, and new growth, mostly birch trees, which can shed radiation annually, has emerged there. The high levels of ionizing radiation have slowed the rate of decay in the forest "because the microbes, fungi, and insects that drive the process of decay also suffered from contamination" (Brown 2019, 128). In the dry, hot summer the forest is prone to wildfires that send up radioactive dust into the air. It is still one of the most contaminated places in the Zone, an area that even workers involved in the power plant's decommissioning avoid.

The fact that the Chornobyl NPP was built on a marshland was a huge mistake. The sandy, swampy marshes along the Prypiat River are particularly good at absorbing radiation, and their seasonal floods contribute to the further spread of radiation: "Swamps … accumulate peaty soils that are rich in organic substances but poor in minerals. Plants searching for potassium, iodine, calcium, and sodium readily take up radioactive strontium, cesium, and iodine that mimic these minerals" (Brown 2019, 134). Radiation then gets into the herbs, berries, and mushrooms that make up the rural diet of humans and animals. In her study of Chornobyl, environmental historian Kate Brown describes the work of biologists Tim Mousseau and Anders Møller who study the ecology of the Zone and "are practicing the kind of science left to post-human landscapes, a science that is tedious and hazardous, as much as it is creative and invigorating" (131). Brown accompanied the pair on several expeditions into the Zone to monitor the diverse flora and fauna. Mousseau and Møller are some of the only biologists who regularly visit the Zone to conduct meticulous fieldwork and observation. What they have found is that the verdant regrowth conceals extensive damage: "The evidence is etched in the ecosystem of the Zone, in the bodies of mice, the leaf litter of the forest floor, and the tumors that cloud the vision of barn swallows they catch" (Brown 2019, 131). Areas with increased radiation also see fewer pollinators, which means that fruit trees produce less fruit and the populations of fruit-eating birds and mammals decline. Similar patterns of damage are evident decades after the disaster and remind us not only of the fragility of ecosystems but also how insidiously these threats manifest. Worse still, science provides us with an incomplete picture so that we are "far from understanding or even being able to register all the radiological effects on the air, water, and soil ecosystems due to anthropogenic radioactive contamination" (Yablokov et al. 2009, 232). Studying the Zone's flora provides valuable information about the postapocalyptic landscape, where plants "teach us that there is no infinite growth, no growth without decay, itself the precondition for future growth" (Marder 2016, 48). We need to direct our attention to the nonhuman perspective on life after the end for insight, because the ecological disruptions evident in the Chornobyl Exclusion Zone are happening elsewhere on larger, more critical scales. Plants follow their own cycles of rebirth and have much to show us about transformation.

The kind of "ecological literacy" that biologists like Mousseau and Møller are devising through careful study and documentation of the local conditions postdisaster is necessary preparation for further apocalyptic scenarios in this epoch of ecological breakdown (Brown 2019, 131). Humans will be dealing with the consequences of nuclear power for thousands of years, not only from the radiation released from Chornobyl but also the legacy of nuclear weapons testing, nuclear plant accidents, and hazardous radioactive waste. Attempts to contain these by-products will only fail, because "nuclear materials, unbounded and unbindable, constantly escape the infrastructures that seek to contain them and forge their own entanglements with us at the cellular and subcellular levels" (Hurley 2020, 205). Given our extensive and dangerous proximity to nuclear materials and the potential for future nuclear disasters, understanding Chornobyl's apocalyptic dimensions may help us "think productively about how to live in the age of nuclear entanglement" (Hurley 2020,

205). Given how unevenly the consequences of nuclear power are felt, the lack of transparency in how the nuclear industry operates, and the dominance of official narratives that may elide certain facts, real experiences of slow nuclear violence are crucial. Ukrainian stalkers, by utilizing unofficial channels to explore the Zone, offer a "counterhistory of apocalypse from below" (Hurley 2020, 38). In contrast to dominant histories of Chornobyl, which often offer a superficial closure to the disaster, they construct a history through personal experience and an intimate knowledge of the landscape that "allows us to glimpse how our societies deal and might deal with future events" (Harper and Specht 2022, 8). The experiences of young Ukrainians such as Markiyan Kamysh and Fedor Oleksandryvych get at the core of how an alternative strategy for knowledge production around the costs of nuclear power might function.

In the Zone, the only legal visitors are tourists, who must be registered with an official tourist agency. The term "stalker" comes from Arkady and Boris Strugatsky's novel *Roadside Picnic* (1972) and was popularized by Tarkovsky's film adaptation *Stalker* (1979) in which stalkers break into the zone created after an alien encounter to steal artifacts. Chornobyl's stalkers are the first generation to grow up after the disaster, and many know or are related to one of the hundreds of thousands of workers and soldiers who were sent to decontaminate the Zone. In *Stalking the Atomic City: Life among the Decadent and Depraved of Chornobyl*, Markiyan Kamysh describes himself as a stalker: "Countless times a year, I am an illegal tourist in the Chornobyl Zone, a stalker, a walker, a tracker, an idiot—you name it. They can't see me, but I am there. I exist. Almost like ionized radiation" (2022, 5). His father was a liquidator who spent six weeks working on the cleanup campaign, but despite the trauma associated with the disaster, Kamysh finds himself constantly drawn to the Zone: "I wanted to sniff and touch every patch of this dump, every fragment of the past. And every time I came back, I swore that it would be my last, my very last visit" (113). Kamysh's attitude toward the Zone is simultaneously irreverent and protective. He is just as likely to call the site of the world's most infamous nuclear disaster a "shitty dump" or "piss-filled Zone" as he is to relish in the Zone's deceptive beauty (112). The combination of irreverence and respect reflects the tension between growth and degradation that characterize the Zone and is most poignantly represented in the contrast between verdant nature and radioactive decay.

Kamysh commands the Zone with a sense of ownership that naturally accompanies a deep understanding of the space. His depth of knowledge cannot be obtained on a casual visit on official tours, which typically follow demarcated routes designed to avoid the most dangerous spots. Stalkers are experts in the Zone's physical topography, as Kamysh acknowledges: "You walk through the bushes that only you know, following the labyrinthine structure of the canals without a map, guided by your nose and sixth sense.

You choose the most distant villages and corners of the Zone" (120–121). From charting their own paths through the overgrown paths and seeking lost villages, they have gathered an extensive knowledge of life there. They know which buildings are safe, where the police patrol, where to drink and forage for food and water, and how to evade wildlife. For instance, Kamysh offers this advice when drinking from the rivers and streams in the Zone:

> I like the taste of the Uzh—bitter but without the metallic taste of the water by the pier in Prypyat. In the river under the railroad bridge, the water tastes a bit better than it does in Starik Creek, but it's not as clear as the water in the Veresnya River—the water in the Veresnya is just delicious. The brook in Kopachi is red, and if you make tea, better splash some Becherovka into it—it will smother the taste of rust.
>
> *(96)*

What Kamysh and other stalkers cultivate through time spent in the Zone is a "situated knowledge" of place that Ursula Heise describes as "the intimate acquaintance with local nature and history that develops with sustained interest in one's immediate surroundings" (2008, 30). Heise explains that the activities of tending to the land and observing the plant and animal life are "some of the ways the human body is perceived to reintegrate itself into the biotic community" (30). The artificial boundary between man and nature must be collapsed to allow for a more ethical and expansive environmental protection strategy. Though reintegrating oneself into a radioactive landscape is not ideal, living in toxic places is becoming the reality for many communities on the frontiers of climate change and industrial pollution. Increasingly, given the extent of globalization, the effects of local environmental disasters are felt globally.

Similarly, Joseph Masco, who writes extensively on American nuclear landscapes, reiterates that it is crucial that we "articulate the ways that planetary-scale fallouts are now differentially remaking local conditions" (2020, 359). Though we know how bodies and environments react to high levels of radiation exposure, persistent exposure to low levels of radiation is not widely studied and may only reveal their damage decades later. The stalkers know they are taking risks by spending so much time in the Zone, but they do not care. Though some might undertake these illegal expeditions purely in search of a thrill, for many stalkers, these illegal visits are about freedom and the ability to control their own fates. Kamysh sardonically jokes that in a few decades, he and other stalkers will meet again in a cancer clinic: "We'll smile at a life that challenges you and dictates where you should walk, how you should live, and what you should breathe. After all, we're the children of our time. Where else could we be?" (2022, 137). They are all too aware of the nuclear risks, in a way that much of the world is free to ignore, because the effects of nuclear disaster are either too remote or invisible. Stalkers do not have the luxury, having been born with the psychological and genetic burdens of a traumatic nuclear legacy. Their mapping of the Zone "employs the political practice of seeing purposefully unmarked and secret landscapes; it makes visible those who have been obscured and silenced within those landscapes" (Kuletz 2016, 7). Tamara Hundorova remarks on this importance, noting that in contrast to the virtual representations of the Zone, in photography, film, and video games, the "stalker stories bring back the reality of the places, dates, and routes in the Chornobyl Zone that were lost or had vanished" (2021, 71). Kamysh also describes his writing about the Zone as a kind of "'geopoetics' in which the dominant element is a place, not a person" (2015). Many stalkers share a similar perspective. Kirylo Stepanets, who has extensively documented the Zone in his book *Chornobyl through the Eyes of a Stalker*, laments the fact that the Zone appears on maps as an "empty space, where many geographic names—have lost meaning from the absence of residents" (2017, 96, translation mine). From the perspective of the natural environment, the absence of human life is beneficial, even preferable, but the emptiness of the Zone, in terms of its visual markers of human habitation, foretells an "end to the possibility of human history" in which "geological time will continue without us, but it will no longer have any meaning without human witnesses to its passage" (Anderson 2016, xviii). The Zone's destabilized hierarchy gives way to a posthuman perspective that forces us to acknowledge our duty of environmental care and responsibility for all living beings or further catastrophe is unavoidable.

Stalkers are acutely aware of the importance of plant and animal life in a posthuman landscape. Their activity and movement in the Zone are often determined by the potential for wildlife encounters, which are more fearsome than an encounter with the police (Kamysh 2022, 127). Kamysh describes one such terrifying encounter with a lynx and her cubs one night after realizing that he and his fellow stalkers were sharing the same abandoned building. At one point the lynx attempted to breach the rotted wooden barricades and the hemmed-in trespassers had to build a fire that might scare her away:

> To the sound of thousands of heartbeats and predatory breathing, you build barricades, stacking doors, while lynx cubs play noisily in the attic and their huge mother lynx breathes heavily, the megahertz of her breathing on kindling Communist pioneer fires of fear and panic.
>
> *(Kamysh 2022, 127)*

Part of stalking involves learning to evade the boars, wolves, and occasional bear, to anticipate their behavior and to read the landscape for signs of suspicious activity. The once solid structures of civilization have deteriorated to become part of the quotidian lives of the nonhuman creatures of the Zone, providing food and shelter: "A bird built a nest on a shelf in the 'red corner' where the icon used to be. Badgers dig their burrows under the floor. Wild pigs, deer, and elk seek out the apples and pears in the wild orchards in the autumn" (Stepanets 2017, 97, translation mine). In these encounters, the nonhuman life of the Zone displaces the human from the center of our cosmic imagination, as it must if there is any possibility of reversing the damage we have already done. The gritty, vulnerable, and dangerous experiences of life in the Zone illuminate a harsh postapocalyptic reality and yet the stalkers, in navigating the risks of this space, have learned out of necessity to coexist with multiple threats and uncertainties. They continually return to the Zone, in a state of ambivalence, in part because they love a place "that has brought so much sorrow to other people" (Stepanets 2017, 322). Gabriele Schwab identifies this search for freedom in the Zone as a sign of nuclear trauma (2020).

The postapocalyptic landscape is also "posttraumatic reality" for many, including the stalkers who were born in the disaster's wake (Hundorova 2021, 71). They are motivated by an urge to probe the past and engage with not only the memory of Chornobyl but also the myths of the Soviet past that created it. They must also reckon with the Zone's current-day mythologization as a dead wasteland. The abandoned buildings and modern ruins are popular visual tropes that have earned Prypiat the descriptor of "Dead City." Documentary films and photography may communicate a misleading version of the Zone that privileges the material trauma and abandonment over the people most acutely affected by Chornobyl. Kamysh offers a scathing assessment of the Zone's tourists, who he feels are responsible for this misrepresentation: "Bored hipsters shot Prypyat dead with their expensive cameras; rich girls from the capital soiled the rotten couches with their tattooed backs and mapped every nook of the terra incognita on Instagram" (2022, 37). Without appropriate context, the eerie photographic tableaux filled with decaying artifacts and empty broken spaces elide the disaster's unfolding reality, fetishizing the disaster while simultaneously reducing its complexity. As beneficial as tourism has been for renewing interest and investment into the Zone, there are limits to its efficacy as a mode of teaching about the disaster.

Kamysh is protective of the Zone not because he is an environmentalist, at least not in the conventional sense, nor from a need to preserve the Zone as a museum. Even as he worries that "crowds of tourists on official tours will snatch up items" and "dismantle the mosaic of beautiful Chornobyl Land," Kamysh frequently mentions tearing down fences to make a fire and smashing windows to enter the abandoned buildings (2022, 62–63). Theirs is a personal connection linked to what Olga Bertelsen terms the Zone's "emotional cartography" as they seek "childhood memories and the secrets of the past" (Bertelsen 2018, 4). By charting territory outside of official tours and narratives, stalkers excavate memory from beneath the detritus and overgrowth, exposing "a reality hidden behind the social world made visible through ruination" (Masco 2020, 242). Theda Wrede writes of the destructive power of the American Southwest's mythologization as simultaneous wasteland and frontier. This myth "reinforced a conceptual division between humans and nature that is at the root of the conquest of the land and the subjugation of the minority populations" (Wrede 2014, 1–2). Similar mythologies were responsible for Ukraine's subjugation under Soviet rule, the trauma of Chornobyl, and Ukraine's current geopolitical precarity, which is why understanding the historical context for the emergence of apocalyptic myth is essential. Too often, myth is discounted as fantastical or abstract, when it may be more productive to collapse the boundary between myth and history. Pablo Richard writes that "even overly cosmic myths express realities in history" and when "cosmic reality is invoked, the aim is to provide drama and urgency to processes within history" (2019, 29).

The historical reality of Chornobyl concerns the "truth about Ukraine's colonial past" (Bertelsen 2018, 25). Ukrainian artist Fedor Oleksandryvych, also has a deep relationship with the Zone, and his obsession with Chornobyl hinges on this colonial inheritance. His family was living in Prypiat during the disaster, and his father was also a liquidator. In the days after the reactor's explosion, Oleksandryvych was evacuated and temporarily placed in an orphanage, a trauma that he processes through his art and in the documentary *The Russian Woodpecker* (Garcia 2015). The film, by Chad Garcia, follows the artist as he investigates a conspiracy theory that the Chornobyl disaster was a government cover-up operation to distract from the failures of Duga, a missile-deterring radar system. The Duga structure is a giant metal grid edifice located within the Zone. Oleksandryvych's claims are not fully substantiated, but the overarching motivation for his theory is compelling, particularly given the Maidan protests playing out alongside his quest for the truth. The political background of Oleksandryvych's project strengthens his primary argument that Russia was and continues to be a major threat to Ukraine's future. By 2014, many young Ukrainians were fed up with the increasing Russian influence and the post-Soviet political culture of corruption and nepotism. Throughout *The Russian Woodpecker*, Oleksandryvych is filmed speaking to crowds about the duplicity of a government that does not protect its citizens. He is deeply suspicious of Russia's motives in interfering with Ukraine's politics, as he should be, considering the legacy of historical trauma inflicted on the country by Russia and the Soviet Union. Roberto Barrios emphasizes the political character of disasters in his work on disaster anthropology, writing that "disasters are perhaps better described as contested arenas where hegemonic visions of societal advancement are challenged by the voices and experiences of those most impacted by catastrophes" (2017, 157). Given Ukraine's contentious and violent relationship and the ecological destruction of war, it is more crucial than ever to understand the historical preconditions of environmental disaster.

In *The Russian Woodpecker*, Fedor Oleksandryvych's argument is that Ukraine's traumatic history originates with the same aggressor—Russia and its Soviet successor. Russia's current-day neo-imperialist reincarnation is its continuation, and the same mechanisms of power responsible for those traumas are controlling Ukraine's current political trajectory, ensuring serious consequences for public health and environmental safety. Chornobyl is a political site for the exploration of not only this traumatic legacy but also the ways in which this historical context brings to light the realities of living within what Ulrich Beck (1992) terms a "risk society." In the irradiated regions of Ukraine and neighboring Belarus, "[t]oxicity and memory meet [...] at both macro- and micro-levels," expanding the concept of ecological damage and its duration (Barcz 2021, 131). In raising the question, *The Russian Woodpecker* opens the space for thinking about what Ukraine's uncertain relationship with nuclear power tells us about its relationship with Russia. Although Ukraine has attempted to distance itself from Russia by aligning itself with Europe, the two countries are still inextricably linked through that Soviet nuclear legacy. Until recently, the fuel used in Ukraine's reactors came from Russia and was processed there, giving Russia undue influence over Ukraine's nuclear operations. Interestingly, the Zone is not a dominant visual feature of Oleksandrovych's film. In fact, much of the footage shows interviews conducted within personal spaces or out in the protest-filled streets of Kyiv. The juxtaposition of striking imagery of the Zone against the domesticated interiors and urgency of legitimate social unrest recalls past struggles and immanent futures of ecological collapse. Whereas Kamysh charts the Zone's more tangible, physical traces, Oleksandrovych rifles through the psychological spaces of disaster and the haunting traumas left by nuclear violence and totalitarianism. Motivated by his personal experience, Oleksandryvych "brings to light the metastases of Soviet totalitarianism, which seemed, if not destroyed, then at least hidden or forgotten" (Kotsarev 2016). In the film, the metastasis becomes evident in the way the older generation is still reluctant to talk about Chornobyl or criticize the Soviet government too explicitly; it is also evident in the corruption of then-President Yanukovych and his parliament that strengthened ties with Russia even as Ukrainians demanded change.

The precarity of the post-Chornobyl political and ecological landscape is made apparent in *The Russian Woodpecker*, for which Oleksandryvych handled the production design. His background in theater helps in his depiction of the disbelief and alienation that accompanies the experience of a nuclear disaster. The camera follows him through various eerie tableaus. In one scene, Oleksandryvych, covered in nothing but plastic, holds a lit torch while swaying through a room of discarded moldy gas masks. In another, he stands on top of the gigantic unnatural Duga, which sits so incongruously in a forest, conjuring up nightmarish Cold War anxieties. His investigation of the Zone and its origins constitutes a deep exploration of the Zone's disquieting apocalyptic dimensions. Artem Ryzhykov's art direction emphasizes the monstrous nature of the Duga Radar Station in sweeping shots of the metal military structure, a visible scar on the landscape. The audacity of mankind's ambitions is on full display, inflicted upon the body of the individual and the natural environment. The stylized, avant-garde scenes juxtaposing the artificiality of the manmade structures with the natural environment remind us of the pervasive role of political power in perpetuating technogenic disasters. Despite the revival of nature in the Zone, as Anna Barcz notes, Chornobyl is a reminder of "how the relationship between humans and the environment fell into despair during an era devoted to building a communist society" (2021, 132). Fedor Oleksandryvych's work on Chornobyl situates the individual within Ukraine's traumatic history, where bodily vulnerability, uncertainty, and ecocide are no longer just apocalyptic omens, but real life.

In *Exposed: Environmental Politics and Pleasures in Posthuman Times*, Stacy Alaimo (2016) draws inspiration from Adriana Petryna's fascinating study of the biopolitics after Chornobyl. Petryna's study examines how those exposed to radiation soon had to leverage the impacts to their health to gain access to key medical resources and social support, which left many disenfranchised and neglected. The concept of exposure provides a productive means of inquiry into the ways in which the individual is denied sovereignty in the Anthropocene when the "exposed subject is always already penetrated by substances and forces that can never be properly accounted for" (Alaimo 2016, 5). Exposure shows just how thoroughly "fundamental boundaries have begun to come undone, unraveled by unknown futures" that, though destabilizing and apocalyptic, can nevertheless point to opportunities to "make sense of the networks of harm and responsibility" and cultivate a new environmental ethics and politics (Alaimo 2016, 2). For Alaimo, the human is displaced to make way for the union of "human with the nonhuman, the inhuman and the more inhuman" through a "sense of immersion within the strange agencies that constitute the world" (12–13). Whereas the radiation exposure willingly welcomed by the Zone's stalkers is a more extreme version of exposure, Alaimo recasts these "entanglements of vulnerability and complicity that radiate from disasters" as the foundation for a kind of environmental activism necessary to restore a strong relationship between humankind and the planet (Alaimo 2016, 5). In this sense, the Ukrainians who frequent the Zone "occupy exposure as insurgent vulnerability [...] to perform material rather than abstract alliances, and to inhabit a fraught sense of political agency that emerges from the perceived loss of boundaries and sovereignty" (Alaimo 2016, 5). Their vulnerability is a radical disavowal of the mechanisms that deny and distance us from the material susceptibility of our ecological predicament. What results from these exposed encounters in the Zone is not a definitive solution to the problems posed by nuclear power or a template for ecological revival, because the Anthropocene is "no time for transcendent, definitive mappings, transparent knowledge systems, or confident epistemologies" (Alaimo 2016, 3). What does result is the need for a more nuanced attention to the local, historically situated complexities of environmental disaster and the protest it inspires in conversation with the "massive political and economic predicaments" that occur across the planet (Alaimo 2016, 10). Additionally, Alaimo insists on the power of creative artistic engagement and practices for shaping our relationship to a world we all share.

For all the trauma and loss associated with the Zone, many stalkers report feeling relaxed there. As time spent in the Zone increases, the "morose feeling of potential death, no longer bothers them" and, instead, "the Zone comforts them and stimulates their acute interest in both life and near death experiences" (Bertelsen 2018, 15). Given the instability of life outside of the Zone and the shadow of an enduring struggle against Russia, the Zone's liminal temporality may feel comforting, even freeing: "Importantly, most stalkers are convinced that Chornobyl liberates people from their fears, making them reject state and nuclear violence and inciting their political activism and longing for freedom" (Bertelsen 2018, 17). Such sentiment was confirmed by stalkers in an interview conducted by Holly Morris, herself a filmmaker of the Zone settler subculture: "Clearly, stalking delivers them the present moment—the lawless wild greenery and pathless reality of the Zone—that makes them, if only for a short while, masters of their destinies" (2014). Acting as documentarians and geographers of this toxic environment, they play an important role in maintaining the disaster's memory, and their experiences communicate vital information about the condition of the Zone. They know the names of villages long forgotten, and their stories and experiences attest to the consequences of nuclear disaster in a way that is not immediately apparent when one looks at the Zone's verdant exteriors. As much as there are inherent risks to spending prolonged time in the Zone, this kind of situated knowledge is "essential to rendering radiological contamination, and other environmental hazards, visible" (Kuchinskaya 2014, 113). Unfortunately, unless we seize the opportunity to imagine and act with the urgency and clarity of vision demanded by the seriousness of our current ecological crisis, we will not be able to recover any stability and hope for a better future.

Conclusion: nuclear power in the shadow of war

In late February of 2022, Russia invaded Ukraine, escalating the war that had started in 2014. Russian forces, entering from Belarus, soon occupied the Zone.[4] Around 300 workers and guards who were maintaining the plant's safety and security were trapped there for the duration of the occupation. A postoccupation assessment found disturbing evidence of a complete disregard by Russian forces for Zone maintenance, resulting in an estimated 54 million dollars' worth of damage (Peter 2022). Equipment was stolen, the plant's laboratories were looted, and military vehicles destroyed roads and bridges going into the Zone. Additionally, trenches were dug in some of the most contaminated areas of the Zone, and plant workers were held hostage for weeks (Peter 2022). Though neither the decommissioned reactor nor the 20,000 spent nuclear fuel rods stored there were damaged, the occupation serves as a warning of the vulnerability of Ukraine's nuclear power plants. This point was further emphasized when the Zaporizhzhia Power Plant, 700 km from Kyiv near the city of Enerhodar, was overtaken by Russian forces after a period of intense fighting that damaged a training laboratory on March 3 (Roth 2022). Russia is now using Europe's largest nuclear power plant as a military base, which poses alarming risks not only for Ukraine but also for Europe (Hinshaw and Parkinson 2022). Russia has also sent engineers from Rosatom, the State Atomic Energy Corporation, to take control of operations. The conditions of war have prevented any external oversight, and communication coming out of the plant is sparse, so should something go wrong in the functioning and safety of the power plant, it is uncertain how efficiently an emergency could be handled. The political connotations of Russia's occupation of the Zaporizhzhia NPP are clear, but the ecological implications are more concerning. The war has already caused serious environmental damage across Ukraine, leading to the displacement of both human and animal populations, disruptions to ecological systems, habitat destruction through bombs, and pollution of natural resources with heavy metals and greenhouse gases, but the possibility of a missile strike on a nuclear facility or nuclear waste storage facility raises nightmarish questions as to nuclear power's inherent vulnerability and the devastating consequences of its failure.

Ironically, the current conflict with Russia has brought more nuclear power to Ukraine. To cut ties with Russia, in June of 2022, the country signed a deal with American energy conglomerate Westinghouse, which will provide nuclear fuel, power plant components, and processing services to Ukrainian NPPs (Digges 2022). This agreement also includes the construction of nine new reactors. Though Ukraine's further commitment to nuclear power in the form of American uranium and new reactors helps Ukraine sever ties to Russia, this move also increases Ukraine's nuclear dependence and risk. This situation can also be understood as an opportunity for the nuclear industry to establish itself as a viable energy solution for the future. We must ask to what extent the current conflict impedes development of a more environmentally conscious future, when before the war, the Zone was transforming into a place of experimentation with greener energy technologies such as The Solar Chernobyl Project.[5] The reversal of post-Chornobyl environmental activism is concerning and unsettling. Such a development only proves a point: that it is too simple to resign Chornobyl to the Soviet past and faulty reactors, because the same infrastructures and goals that caused the nuclear disaster continue to decide our future. This point is echoed by Plokhy, who ominously warns that "there is no doubt that a new Chernobyl-type disaster is more likely to happen if we do not learn the lessons of the one that has already occurred"

(2018, xvi). With Ukraine's decision, the anti-nuclear sentiments of the post-Chornobyl era have given way to a renewed call for modernization, hearkening back to early Soviet nuclear culture utopian myths. Once again, political and economic concerns are privileged above environmental ones despite clear drawbacks. Ukraine's renewed reliance on nuclear power "is a reminder of the hypocrisy of learning to live with apocalypse" (Barcz 2021, 144).

The importance of mythological thinking in conceptualizing the apocalyptic dimensions of nuclear disaster will only become more prevalent as we confront the Anthropocene crises that threaten all life on Earth. Unfortunately, environmental harm will only increase unless we realize how destructive our tendencies to think of the Earth as an afterthought and not the most essential key to our survival. In a cruel, seemingly inevitable twist, the war in Ukraine has reconfirmed our investment in nuclear power and muted any warnings to do otherwise. In the *Chernobyl Herbarium* (2016), philosopher Michael Marder, who was exposed to radiation in Anapa, Russia as a boy, reflects on his Chornobyl connections alongside Anais Tondeur's ethereal photograms of Zone plants. He suggests that mankind can no longer stand as an ethical center and environmental caretaker: "Instead of being the masters of our milieu, we are lost on a planet transformed and mutilated because of human activity. Worse still, the internal compass, which was our consciousness, is shattered and no longer usable" (48). As unforgiving as the possibility may be, we are living in a state of alarming environmental crisis, and the "frequency, severity, and magnitude of disasters is increasing, and [...] we are making our environment and societies more and more vulnerable to catastrophe" (Button 2010, 249). Each apocalyptic event might offer its solutions and revelations, but any guidance will cease to matter unless we are willing to see how unprepared we are for the environmental realities awaiting us.

What is needed is ecological literacy, one that prepares us for postapocalyptic living while it also pushes us to creative action and alternative modes of knowledge production. Failure to consider the environmental repercussions and what it means to inhabit spaces of apocalypse guarantees further environmental harm. The cyclical nature of apocalypse does not mean we are impervious to its dangers. Examining this postdisaster space "anticipate[s] the ecocentric ways in which post-Chernobyl trauma could be overcome, or at least redirected into creating a new environmental history of regions affected by fallout" (Barcz 2021, 132). Apocalyptic thinking ushers in a new environmentalism for the Anthropocene that eschews idealism to focus "on the qualities, consequences, and insights of living in messy aftermaths" (Masco 2020, 263–264). Stalkers "make dead cities alive again" and prompt us to think about living through an apocalypse (Kamysh 2022, 135). Kamysh and Oleksandryvych's Zone explorations are different, but each underscores the necessity of challenging the structures of political power and control that encourage and shape our relationship to nuclear energy. It is imperative that we confront the consequences of our pursuit of technological progress and interrogate the origins of the ecological harm of the nuclear industry that we have inflicted on the planet, especially because "radioactive fallouts affecting large areas and hundreds of thousands of people are likely to reoccur in the foreseeable future," as the Fukushima nuclear disaster demonstrated in 2011 (Shkaruba and Skryhan 2019, 152). Additionally, the nuclear legacy and current war in Ukraine combat the destructive myth that nuclear power is either safe or a panacea to the climate crises. Unless we can seize the opportunity to imagine an alternative future and divest from nuclear power, apocalyptic events will be a reality to which we have to adapt.

Notes

1 In keeping with recent guidelines established by the Ukrainian government, Ukrainian spellings will be used ("Chornobyl" instead of "Chernobyl"). However, some quotations may still contain Russian spellings.
2 In this discussion, "mythology" works on multiple levels, referring to not only the biblical mythology of end-times but also the mythologies of everyday life.
3 Though the disaster became an immediate priority in 1991 when Belarus gained its independence, the economic and public health burdens connected with the disaster were gradually displaced as the Belarusian government moved to control the production of knowledge around the disaster and its consequences. See Kuchinskaya (2014) and Shkaruba and Skryhan (2019).
4 This chapter was written during the summer of 2022.
5 https://solarchernobyl.com.

References

Alaimo, Stacy. 2016. *Exposed: Environmental Politics and Pleasures in Posthuman Times.* Minneapolis: University of Minnesota Press.
Alexievich, Svetlana. 2006. *Voices from Chernobyl: The Oral History of a Nuclear Disaster*, translated by Keith Gessen. New York: Picador.
Anderson, Mark. 2016. "Introduction: The Dimensions of Crisis." In *Ecological Crisis and Cultural Representation in Latin America: Ecocritical Perspectives on Art, Film, and Literature*, edited by Mark Anderson and Zelia M. Bora, xi–xxxii. Lanham, MD: Lexington Books.
Barcz, Anna. 2021. *Environmental Cultures in Soviet East Europe: Literature, History and Memory.* London: Bloomsbury Academic.
Barrios, Roberto E. 2017. "What Does Catastrophe Reveal for Whom? The Anthropology of Crises and Disasters at the Onset of the Anthropocene." *Annual Review of Anthropology* 46: 151–166. https://doi.org/10.1146/annurev-anthro-102116-041635.
Beck, Ulrich. 1992. *Risk Society: Towards a New Modernity*, translated by Mark Ritter. London: Sage.
Berezhnaya, Lilya. 2018. "Ukrainian Millennialism: A Historical Overview." In *The Apocalyptic Complex: Perspectives, Histories, Persistence*, edited by Nadia Al-Bagdadi, David Marno, and Matthias Riedl, 253–280. Budapest: Central European University Press.
Bertelsen, Olga. 2018. "Chornobyl as an Open Air Museum: A Polysemic Exploration of Power and Inner Self." *Kyiv-Mohyla Humanities Journal* 5: 1–35. https://doi.org/10.18523/kmhj150381.2018-5.1-36.

Brown, Kate. 2019. *Manual for Survival: A Chernobyl Guide to the Future*. New York: W.W. Norton & Company.

Bruno, Andy. 2016. *The Nature of Soviet Power: An Arctic Environmental History*. New York: Cambridge University Press.

Button, Gregory. 2010. *Disaster Culture: Knowledge and Uncertainty in the Wake of Human and Environmental Catastrophe*. Walnut Creek, CA: Left Coast Press.

Digges, Charles. 2022. "Westinghouse to Supply All of Ukraine's Nuclear Fuel," *Bellona*, June 3, 2022. Accessed July 26, 2022. https://bellona.org/news/nuclear-issues/2022-06-westinghouse-to-supply-all-of-ukraines-nuclear-fuel.

Drach, Ivan. 1974. Корінь і крона: Поезії [The root and the crown: Poetry]. Kyiv: Radyanskyi Pysmennyk.

Feshbach, Murray, and Alfred Friendly, Jr. 1992. *Ecocide in the USSR: Health and Nature Under Siege*. New York: Basic Books.

Gracia, Chad, dir. 2015. *The Russian Woodpecker*. Roast Beef Productions, Rattapallax, and Gracia Films.

Harper, Earl T., and Doug Specht. 2022. "Introduction: … these unprecedented times." In *Imagining Apocalyptic Politics in the Anthropocene*, edited by Earl T. Harper and Doug Specht, 1–14. Abingdon, UK: Routledge.

Heise, Ursula K. 2008. *Sense of Place and Sense of Planet: The Environmental Imagination of the Global*. Oxford: Oxford University Press.

Hinshaw, Drew, and Joe Parkinson. 2022. "Russian Army Turns Ukraine's Largest Nuclear Plant into a Military Base," *Wall Street Journal*, July 5, 2022. https://www.wsj.com/articles/russian-army-turns-ukraines-largest-nuclear-plant-into-a-military-base-11657035694.

Hundorova, Tamara. 2019. *The Post-Chornobyl Library: Ukrainian Postmodernism of the 1990s*, translated by Sergiy Yakovenko. Brookline, MA: Academic Studies Press.

Hundorova, Tamara. 2021. "Post-Chornobyl: From (Non)Representation to an Ecocritical Reading of Nuclear Trauma." *Slavia Iaponica* 24: 59–75.

Hurley, Jessica. 2020. *Infrastructures of Apocalypse: American Literature and the Nuclear Complex*. Minneapolis: University of Minnesota Press.

Josephson, Paul R. 1996. "Atomic-Powered Communism: Nuclear Culture in the Postwar USSR." *Slavic Review* 55, no. 2: 297–324. https://doi.org/10.2307/2501914.

Josephson, Paul R. 2005. *Red Atom: Russia's Nuclear Power Program from Stalin to Today*. Pittsburgh: University of Pittsburgh Press.

Josephson, Paul R., Nicolai Dronin, Aleh Cherp, Ruben Mnatsakanian, Dmitry Efremenko, and Vladislav Larin. 2013. *An Environmental History of Russia*. Cambridge: Cambridge University Press.

Kamysh, Markiyan. 2015. "Маркіян Камиш: Плавання серед чорнобильських боліт було чимось на зразок колумбової мандрівки." [Markiyan Kamysh: Swimming through the Chornobyl Swamps was like Columbus' Journey]. Interview by Irina Chervinska. *Vikna*, December 21, 2015. https://vikna.if.ua/news/category/articles/2015/12/21/46336/view.

Kamysh, Markiyan. 2022. *Stalking the Atomic City: Life among the Decadent and the Depraved of Chornobyl*, translated by Hanna Leliv and Reilly Costigan-Humes. New York: Astra Publishing House.

Kotsarev, Oleg. 2016. "Самопрезентація совка. Фільм Російський дятел не так про Чорнобиль, як про радянське минуле." [Self-presentation scoop. The film *The Russian Woodpecker* is not so much about Chernobyl as about the Soviet past]. Texty.org.ua, February 19, 2016. https://texty.org.ua/articles/65405/Samoprezentacija_sovka_Film_Rosijskyj_datel_ne_tak-65405/.

Kuchinskaya, Olga. 2014. *The Politics of Invisibility: Public Knowledge about Radiation Health Effects after Chernobyl.* Cambridge, MA: MIT Press.
Kuletz, Valerie. 2016. *The Tainted Landscape: Environmental and Social Ruin in the American West.* London: Routledge.
Lindbladh, Johanna. 2019. "Representations of the Chernobyl Catastrophe in Soviet and Post-Soviet Cinema: The Narratives of Apocalypse." *Studies in Eastern European Cinema* 10, no. 3: 240–256.
Marder, Michael. 2016. *Chernobyl Herbarium: A Fragments of an Exploded Consciousness*, with artworks by Anais Tondeur. London: Open Humanities Press.
Masco, Joseph. 2020. *The Future of Fallout, and Other Episodes in Radioactive World-Making*. Durham: Duke University Press.
Morris, Holly. 2014. "The Stalkers," *Slate*, September 26, 2014. https://slate.com/news-and-politics/2014/09/the-stalkers-inside-the-youth-subculture-that-explores-chernobyls-dead-zone.html.
Mycio, Mary. 2005. *Wormwood Forest: A Natural History of Chernobyl.* Washington, DC: Joseph Henry Press.
Mykhailenko, Anatoliy. 1988. "Запах полину. Повість—пам'ять" [The smell of wormwood. A story—a memory]. *Dnipro* 62: 4–57.
Mykoliuk, Oksana. 2009. "Seeking Place for Polissian Atlantis," *Den'*, April 28, 2009, https://m.day.kyiv.ua/en/article/day-after-day/seeking-place-polissian-atlantis.
New International Version. 2011. "Biblica." https://www.biblegateway.com/passage/?search=Revelation%208&version=NIV.
Nixon, Rob. 2011. *Slow Violence and the Environmentalism of the Poor.* Cambridge, MA: Harvard University Press.
Oliinyk, Borys. 1988. *Address to Vsesoiuznaia Konferentsiia Kommunisticheskoi Partii Sovetskogo Soiuza, June 28–July 1, 1988*. Moskva: Izdatel'stvo Politicheskoi Literaturi.
Onyshkevych, Larissa M.L. Zaleska. 1990. "Echoes of Chornobyl in Soviet Ukrainian Literature," *Agni*, no. 29/30: 279–291.
Oppo, Andrea. 2013. *Shapes of Apocalypse: Arts and Philosophy in Slavic Thought.* Boston: Academic Studies Press. https://doi.org/10.2307/j.ctt1zxshsh.
Peter, Laurence. 2022. "Ukraine War: Chernobyl Scarred by Russian Troops' Damage and Looting." BBC News, June 3, 2022. https://www.bbc.com/news/world-europe-61685643.
Phillips, Sarah D. 2008. "Chernobyl's Sixth Sense: The Symbolism of an Ever-Present Awareness." *Anthropology and Humanism* 29, no. 2: 159–185. https://doi.org/10.1525/ahu.2004.29.2.159.
Plokhy, Serhii. 2018. *Chernobyl: The History of a Nuclear Catastrophe.* New York: Basic Books.
Richard, Pablo. 2009. *Apocalypse: A People's Commentary on the Book of Revelation.* Eugene, OR: Wipf & Stock.
Riedl, Matthias, and David Marno. 2018. "The Resilience of the Apocalyptic." In *The Apocalyptic Complex: Perspectives, Histories, Persistence*, edited by Nadia Al-Bagdadi, David Marno, and Matthias Riedl, vii–xxi. Budapest: Central European University Press.
Romashko, Elena. 2016. "Religion and 'Radiation Culture': Spirituality in a Post-Chernobyl World," *Material Religions Blog*, June 1, 2016. https://materialreligions.blogspot.com/2016/05/religion-and-radiation-culture.html.
Romashko, Elena. 2019. "Russian Orthodox Icons of Chernobyl as Visual Narratives About Women at the Center of Nuclear Disaster." In *Orthodox Christianity and*

Gender: Dynamics of Tradition, Culture and Lived Practice, edited by Helena Kupari and Elina Vuola, 190–209. Abingdon, UK: Routledge.

Roth, Nickolas. 2022. "What Happened at Ukraine's Zaporizhzhia Nuclear Power Plant and What are the Implications?" *Nuclear Threat Initiative*, March 9, 2022. https://www.nti.org/atomic-pulse/what-happened-at-ukraines-zaporizhzhia-nuclear-power-plant-and-what-are-the-implications/.

Schmid, Sonja D. 2015. *Producing Power: The Pre-Chernobyl History of the Soviet Nuclear Industry*. Cambridge, MA: MIT Press.

Schwab, Gabriele. 2020. *Radioactive Ghosts*. Minneapolis: University of Minnesota Press.

Shevchuk, Vasyl Yakovych, Yuriy Mykolayovych Satalkin, Georgiy Oleksiyovich Bilyavsky, Vasyl Mykolayovych Navrotsky, and Oleksandr Oleksandrovych Mazurkevich. 2005. *Preserving the Dnipro River: Harmony, History and Rehabilitation*. Oakville, ON: Mosaic.

Shkaruba, Anton, and Hanna Skryhan. 2019. "Chernobyl Science and Politics in Belarus: The Challenges of Post-normal Science and Political Transition as a Context for Science–Policy Interfacing." *Environmental Science & Policy* 92: 152–160. https://doi.org/10.1016/j.envsci.2018.11.024.

Stepanets, Kyrylo. 2017. *Чернобыльская зона глазами сталкера* [The Chornobyl Zone through the eyes of a stalker]. Kyiv: Sky Horse Publishing.

Strugatsky, Arkaday, and Boris Strugatsky. 1972. *Пикник на обочине* [Roadside picnic]. Moscow: Molodaya Gvardia.

Tarkovsky, Andrei, dir. 1979. *Сталкер* [Stalker]. Mosfilm, DVD.

Whiteley, John M. 1999. "The Compelling Realities of Mayak." In *Critical Masses: Citizens, Nuclear Weapons Production, and Environmental Destruction in the United States and Russia*, edited by Russel J. Dalton, Paula Garb, Nicholas P. Lovrich, John C. Pierce, and John M. Whiteley, 59–96. Cambridge, MA: MIT Press.

Wrede, Theda. 2014. *Myth and Environment in Recent Southwestern Literature: Healing Narratives*. Lanham, MD: Lexington Books.

Yablokov, Alexey V., Vassily B. Nesterenko, and Alexey V. Nesterenko. 2009. "Atmospheric, Water, and Soil Contamination after Chernobyl." In *Chernobyl: Consequences of the Catastrophe for People and the Environment*, 223–236. Boston: Blackwell.

Yavorivskyi, Volodymyr. 1987. "Марія з полином при кінці століття" [Maria with the bitter wormwood at the end of the century]. *Vitchyzna* 7: 16–139.

2

MYTHS OF WILDERNESS AND MOTHERHOOD IN POSTAPOCALYPTIC NARRATIVES OF THE ANTHROPOCENE

Hope Jennings and Christine Junker

"Mother earth" and wilderness as refuge

In two recent postapocalyptic narratives of the Anthropocene, Claire Vaye Watkins' *Gold Fame Citrus* (2015) and Diane Cook's *The New Wilderness* (2020), the theme of maternal abandonment is directly linked to ecological catastrophe and grief. Confronted with a world on the brink of human extinction, where there seems to be no hope of survival for their children, the very concept of motherhood as a source of sustenance and protection appears untenable to the mothers in the texts, Bea and Luz, who choose to fail as mothers. However, as characters struggling to live in the contexts of a damaged world and environmental precarity, Bea and Luz are only failures as mothers based on conventional assumptions in postapocalyptic Anthropocene narratives that the central objective is and should be human survival. Their failure to live up to the myth of the "good" mother—an endlessly self-sacrificing, nurturing, sustaining source of life—is deliberately framed within the ecological failure of "mother earth" to function as an available resource solely dedicated to the survival of human life. Both novels provide narratives centered on myths of wilderness and motherhood as conflated spaces of refuge only to disrupt and unravel how these myths are unsustainable in the face of escalating climate emergencies and attendant forms of grief and despair over the possibility of survival.

Themes of motherhood, survival, and environmental precarity are prevalent in earlier short story collections by Cook and Watkins, who both explore the centrality of wilderness myths and spaces in the American imaginary. In *Battleborn* (2012), Watkins highlights the history and problematic mythology of the "West," and the complicated nature of motherhood emerges in nearly every one of the stories as well as in her autobiographical novel, *I*

DOI: 10.4324/9781003348535-4

Love You but I've Chosen Darkness (2021). *Gold Fame Citrus*, Watkins' debut novel, expands her critique of American myths of the frontier alongside patriarchal myths of the "good" mother. Likewise, Cook's *The New Wilderness* is her first novel and picks up on familiar themes found in her dystopian collection *Man v. Nature* (2015), where many of the stories emphasize the brutal realities of wilderness spaces in tension with human struggles for survival. In the story "Somebody's Baby," Cook explicitly grapples with the fears and crushing demands of early motherhood, anxieties that plague Bea throughout *The New Wilderness* and are amplified to the point where Bea abandons her young daughter, much like Luz in *Gold Fame Citrus.* Both women come up against the limits and impossibilities of fulfilling traditional roles of motherhood in a world determined by ecological collapse and mass species extinction.

Through their novels' emphases on maternal ambivalence and abandonment, Cook and Watkins offer extended opportunities to critique patriarchal myths of motherhood as an eternal, universal, nurturing force alongside gendered mythologies of wilderness, or "mother earth," as a source of freedom, refuge, and spiritual renewal. Neither wilderness nor motherhood have ever been that simple; both denote a material reality—as actual spaces or lived experiences—and a discursive construction of that reality through cultural stories and myths about wilderness spaces and what it means to be a mother. *Gold Fame Citrus* and *The New Wilderness* attempt to complicate myths of motherhood and wilderness by drawing attention to how these are entangled in the American imagination and exposing how what we take to be a given reality is far more often a mythic construction that privileges the dominant culture.

The American wilderness functions as a potent mythic space in the country's imagination and settler history. From the inception of the nation's colonial enterprise, wilderness represented a terrifying spiritual and material threat to the survival and well-being of the Puritans (Nash [1967] 2014). This attitude persisted into the westward settlement of the frontier but became intertwined with a counterimpulse to idealize wilderness as a space that allowed an escape from the demands of civilization. Through the influence of late 19th-century conservation movements, the wilderness was further transformed from a place of perceived danger and necessary domestication to an endangered, vanishing space (due to colonial expansionism) that needed to be preserved as a source of refuge and freedom from the restraints and illnesses of modern sub/urban life (Theriault and Mowatt 2020). As a mythic place, the wilderness is an unstable locus of revision and reinterpretation, and access to the wilderness as a site of conquest and refuge for the beneficiaries of settler colonialism has rarely been afforded to nonwhite or nondominant Americans (Finney 2014; Theriault and Mowatt 2020).

Myths of motherhood likewise function to perpetuate the dominant culture, specifically patriarchy, which defines motherhood as the primary desire and natural role of all women and scripts the myth of the "good" mother as selflessly nurturing, giving, and dedicated to the well-being of others (Johnston and Swanson 2003). As Deirdre Johnston and Debra Swanson note, feminist scholars consistently point out that the myth of motherhood as "the joyful fruition of every woman's aspirations" is a pervasive patriarchal discourse that "attribute[es] any maternal unhappiness and dissatisfaction to failure of the mother" (22).[1] Thus the unredeemable shadow of the "good" mother: the "bad" mother, the "unnatural" or "monstrous" mother, the mother who neglects, abuses, or abandons her children and who serves as a warning to any woman whose behavior transgresses the prescriptive functions of motherhood.[2] Like the juxtaposed myths of "good" and "bad" mothers, the myth of "mother earth," according to feminist psychotherapist Jan Baker (2013), provides a binary personification of nature as "a receptive and loving source" of maternal nourishment and abundance *and* "a terrifying implacable force [...] eternally indifferent to our fate" (55). In other words, the myth of "mother earth" is based on expectations that mothers and nonhuman nature *can* and *should* provide everything needed for survival and sustenance set against conflicting fears that they might refuse or fail to provide or threaten our safety in ways beyond our control. This psychological ambivalence also underscores settler colonial conflations of earth-as-mother and discloses the complicated, contradictory, and often destructive mythologies of wilderness and motherhood.

It is important to understand how the American myth of wilderness functions as both a settler colonial and gendered place of the imagination that is associated with white masculinity and pioneering independence while the wilderness itself is feminized through the myth of a limitlessly bountiful "mother earth" (Nash [1967] 2014; Comer 1997). This patriarchal settler fantasy of the wilderness-as-mother in terms of property ownership and resource extraction has had devastating ecological consequences. Annette Kolodny (1973) explains in her early ecofeminist essay "The Land-as-Woman" how the myth of wilderness in terms of maternal abundance was also driven by fear of too much bounty or excess. To counter this excess, wilderness needed to be controlled lest settlers, or "civilized" men, regressed to a state of "savagery" (172–173). By the late 19th century, once the national agenda of manifest destiny had been fulfilled and the frontier firmly settled, the wilderness fantasy became one of loss and guilt; it no longer offered the promise of "harmonious return" to nature but "the spectre of destruction and exploitation" due to three centuries of "single-minded cultivation, civilization and mastery of the maternal [...] continent" (178–179). The consequences of settler colonial and capitalist consumption of wilderness resources prompted urgent demands for the protection of "mother earth" and wilderness spaces. After all, who would Americans be without wilderness, which gave birth to the frontier mythology at the heart of (white, male, able-bodied) American identity (Nash [1967] 2014)?

The early 20th-century wilderness preservationist movement was led by nature writer and conservationist John Muir, who argued that protecting federally designated wilderness spaces served the social, emotional, and psychological needs of (white) Americans and their future generations. In *Our National Parks* (1901), Muir explicitly frames the "love of wild nature" as a universal expression of "ancient mother-love" (cited in Nash [1967] 2014, 128). The creation of a National Parks System would protect and preserve "mother earth," who in return would provide "thousands of tired, nerve-shaken, over-civilized people" the chance to become replenished by the natural "fountains of life" found in her remote mountains, valleys, and forests (140). Muir's view of wilderness as a site of spiritual and maternal sustenance, "as a distant, pure nature," had a lasting influence on wilderness myths, if not the actual uses and abuses of wilderness resources (Sutter 2021, 597). Despite the passage of the 1964 Wilderness Act, intended to preserve areas "where the earth and its community of life are untrammeled by man, where man himself is a visitor who does not remain," the notion of wilderness as a space "untrammeled" by humans seems increasingly irrelevant due to the extensive anthropogenic effects of global climate change (Wohl 2013, 4). It is precisely within these historical contexts of the American environmentalist movement and the current climate crises of the Anthropocene that Watkins and Cook frame their representations of wilderness spaces and myths.

The next section of this chapter examines how myths of "mother earth" and wilderness as refuge are entangled in the American cult of wilderness and contemporary ecofascist survivalist fantasies. *Gold Fame Citrus* and *The New Wilderness* play out wilderness survivalist plots, in which the return to wilderness is seen as a return to nature, or "mother earth," and the possibility of salvation for the very people who destroyed the health of the planet. Watkins and Cook, however, expose the patriarchal settler ideologies underpinning the myth of wilderness as benevolent mother. In both novels, wilderness is represented as brutal and inhumanly indifferent, and the failure of mothers and "mother earth" to protect their children suggests that survival in the face of extinction is not going to emerge based on past or current mythologies. Instead, as the latter half of the chapter argues, we might read Bea's and Luz's choices to abandon their daughters as acts of resistance and refusal to adhere to the old myths when confronted with a world that will not and cannot sustain wilderness spaces or traditional modes and practices of mothering. Reading the theme of maternal abandonment through the lens of Rosi Braidotti's nomadic subjectivity, we argue that Luz and Bea attempt to meet the demands of a posthuman shift in human–nature relations and thus reach a different understanding of the terms of human survival in the contexts of the Anthropocene. According to Braidotti (2011) and Stacy Alaimo (2016), and building upon Aldo Leopold's land ethic, those terms of survival require a renewed wilderness ethic, wherein human and ecological entanglements are sustained through vital practices of mutualism and care; ongoing investments in restoration and rewilding; and ecocentered narratives or myths that imagine the possibility of a livable future for humans and nonhumans alike.

Survival and the cult of wilderness

The New Wilderness and *Gold Fame Citrus* deconstruct and critique wilderness survival myths where "mother earth" is imagined as the site of salvation for only the privileged few. Both novels appear to follow what Krista Comer (1997) calls the "wilderness plot," which emphasizes the idea that wilderness offers a "redemptive" and "utopic possibility" (75, 79). Bea and Luz participate, albeit skeptically, with the fantasy of wilderness as the last available place that might provide deliverance from the dystopic consequences of climate collapse and urban deterioration, or at least a chance at survival for their children. However, Watkins and Cook critically dismantle the wilderness plot—where a return to nature or "mother earth" promises salvation for a select community—through their focus on the American cult of wilderness, ecofascist survivalist narratives, and problematic ideas about human–nature relations within some of the more radical strands of modern environmental discourses.

The cult of wilderness took hold of the American imagination in the late 19th century alongside a "sense of discontent with civilization" and fears of American decline (Nash [1967] 2014, 145). Like the preservationist movement, the cult of wilderness associated the frontier "with sacred American virtues" rooted in the "pioneer past" but primarily "as a source of virility, toughness, and savagery—qualities that defined toughness in Darwinian terms" (146, 145). The emulation and worship of a patriarchal frontier masculinity have had an enduring influence on right-wing survivalist narratives that promote the nation's settler colonial past as a foundation of American exceptionalism. According to Johannes Kaminski (2021), fringe survivalist fantasies employ the trope of wilderness as a space to envision a "better" world: a future "post-collapse America" where the wilderness setting functions "as a playground for true masculinity" and a white-supremacist nostalgic return to America's mythic frontier (5, 7). The core desires within these survivalist narratives are less about anxiety over a precarious dystopian future but more invested in an ecofascist utopia where only the (predetermined) fittest deserve to survive.

The New Wilderness deconstructs the cult of wilderness as a settler colonial fantasy to expose and critique contemporary ecofascist survival politics. The novel sets up the traditional wilderness plot, in which the wilderness functions as a romanticized alternative to urban spaces, while building upon the premise of a postapocalyptic wilderness fantasy grounded in a return to tribal primitivism as a solution to American decline. Bea volunteers to live in the "Wilderness State" so that her daughter, Agnes, might escape the toxic pollution of a near-future, overpopulated "City." Bea is wary of the "utopic possibility" of the Wilderness State yet sees no alternative if she wants to save her daughter "because there was no other place they could go" (Cook 2020, 51). The Wilderness State offers the only possibility of urban flight aside from the "Private Lands," which are remote and restricted enclaves for the billionaire class.

The Wilderness State is a refuge, then, though not in Muir's spiritual sense but literally. It is one of the last remaining federally protected lands somewhere in the American west or southwest and promoted as an "untrammeled" wildlife refuge, where a small group of human volunteers are selected to dwell as a nomadic, primitivist "Community" with "Rangers" monitoring and assessing their impact on nature (Cook 2020, 18–19). Human presence and interaction within the Wilderness State is regulated with strict rules that parody the "Leave No Trace" movement of the 1970s and its influence on wilderness tourism and conservation. Like the outdoor enthusiasts of the mid-20th century who were pouring into the National Parks, trampling flora, and disturbing wildlife, the Community members are presumed, just by their presence, to have a negative impact on the environment and must adhere to their Leave No Trace handbook or risk eviction. For instance, they must dispose of all human waste according to stringent guidelines, they may not stay in the same place more than seven nights, and they are not allowed to grow food, build structures, or domesticate animals. In other words, they are prohibited from becoming an agricultural society and must remain nomadic hunters and gatherers, with the assumption that settled, agricultural life inevitably ruins the environment.

This view of a "genuine" or pure wilderness indicates a failure to understand how "humans are natural beings interacting with their natural environment" (Zimmerman 1995, 232). A key tension in the novel is the insistence that preserving "wilderness" is fundamentally incompatible with human needs for familiarity, as if humans cannot be at home in nature or interact with nature in mutually beneficial ways. At one point, Bea's daughter Agnes observes her own discomfort with the Leave No Trace rules when she realizes they are intended to alienate the Community from developing a deep familiarity with the land by destroying "their sense of home" (232). Agnes rightly senses that the Wilderness State is an impossible ideal to uphold because it is impossible for humans to exist without also leaving traces of that existence. No matter how much they try to reduce their impact, humans are no more separate from nature or their environment than any nonhuman presence (or absence) that affects the ecology of a place (Alaimo 2008). The Leave No Trace philosophy in Cook's novel thus asserts a division between human and nonhuman nature when no such separation exists.

The Leave No Trace policies of the Wilderness State are also mired in an ecofascist response to the Anthropocene and the climate crisis. Coined by Michael E. Zimmerman (1995) in his essay, "The Threat of Ecofascism," the term "ecofascism" is linked to far-right radical strands of the environmentalist movement that assert tribal primitivism as the only way to "merge the human and natural," thus promoting "fantasies about plunging back into 'nature' in a way that surrender[s] moral responsibility and historical self-consciousness" (234). Ecofascist wilderness survivalist fantasies are equally invested in white settler myths of the American wilderness as a pristinely untouched space prior to European settlement. Kylie Crane (2012) points out how this myth of "wilderness entails a colonial gesture" that refuses to account for the historical presence of Indigenous populations and their own interactions with the land (2). According to Crane, not only do settler and postsettler myths of wilderness erase Native presences and sovereignty, in the contexts of radical environmentalist and ecofascist discourses such myths assert a settler colonial view of relations between humans and nature as fundamentally separate and destructive, a view that has no equivalency in Indigenous myths and practices. In other words, as Zimmerman argues, ecofascist fantasies of wilderness as a place of primitivism and white-supremacist nostalgia violently expunge other philosophies, histories, or practices of human-nature interactions that reduce "ecological destruction while providing for human well-being" (1995, 232). The cult of wilderness in Cook's novel therefore represents a postapocalyptic survival fantasy that cannot guarantee refuge from the climate crisis because it refuses to imagine a future (or past) where human and nonhuman nature might interact and thrive through mutually beneficial relations and where nature is no longer exploited as an endless resource for the privileged few. As we discover at the end of *The New Wilderness*, the Wilderness Sate is redeveloped into Private Lands and was perhaps always solely intended as a refuge for the rich.

Gold Fame Citrus offers a similar wilderness plot of returning to "pure" nature as an antidote to urban blight. The novel is set in a near-future California decimated by environmental collapse after a generation of drought and wildfires. Though Luz and her partner Ray seem to have reconciled themselves to this postapocalyptic world—aimlessly squatting and scavenging in an abandoned Laurel Canyon house—when they kidnap Ig, a toddler of uncertain origins, they are convinced that they need to find somewhere more habitable to raise their new daughter, whom they believe they have saved from neglect and abuse. They decide to seek out an alternative in the ever-growing Amargosa Dune Sea (located in the Mohave Desert), following rumors of a community led by a water-dowsing "prophet." In a move that hearkens back to the frontier settler mentality that caused the very inhabitability of the "West" they are now living in, the newly formed family sets off for this mystical haven and the imagined possibility of an unknown space they believe will be better than what they have now.

Luz, however, recognizes the fallibility of the frontier myth before they depart: "There was always a savior out in the wilderness. [...] So familiar, this stagey faith" (Watkins 2015, 72). Her skepticism undercuts the wilderness plot because Luz is under no illusion that there is a "savior" in the form of a single individual or institutional solution to the climate crisis, and the idea of a wilderness redemption is "stagey" at best—only a "familiar," outworn myth. Like Bea, Luz turns to the wilderness because she believes there is no other option. Although she resists the myth of wilderness as "mother earth," Luz is seduced by the Amargosan prophet, Levi, and his visions of the expanding dune wasteland as utopic Eden, which Levi documents in his "primer," the sacred text of his wilderness cult and filled with fanciful illustrations and descriptions of the desert flora and fauna that "turned a world once shriveled into a locus of succor" (202). Luz knows this wilderness garden is imaginary, a product of drugged hallucinations, but the primer offers hope for a renewed world where survival is meaningful, communal, life-sustaining, and not prioritized through self-interested competition and rapacious consumption of nature (212).

Levi's wilderness cult appears to reject ecofascist wilderness survival fantasies in favor of a "progressive" environmentalism that promotes the return to wilderness, or "mother earth," as an opportunity for people to exist in harmony with nature. Living beyond the margins of society, sharing resources, and following the shifting sands of the unmapped Amargosa, the caravan commune led by Levi believes they are leading a more "authentic" life. They are New Age frontier pioneers, and their belief is based on the premise that freedom from capitalism and technology—all the trappings of modernity—implicitly instills a better, nonhierarchical way of seeing human and nonhuman nature as inextricably intertwined. In the end, though, Levi is nothing more than a self-appointed delusional wilderness savior. His frontier mythology offers hokey visions of a transcendental oneness with "divine" nature through "communion with the rock and the silt" (Watkins 2015, 153), while insisting on a messianic purpose of being "chosen" by the dune (160–161). Although Levi's visions offer an alternative to hopelessness, the wilderness remains a privileged mythical space for those who are enlightened enough to accept "a seismic shift in the way we understand the environment … blending of the spiritual and the natural" (284). That "shift" is simply a rehashing of the wilderness plot: no matter how badly they have abused "mother earth," humans can return to the wilderness for spiritual sustenance and survival, and the wilderness will be the benevolent "good" mother, healing and nurturing her chosen children.

From the beginning of *Gold Fame Citrus*, though, "Mother Nature" is no longer functioning like a "good," accommodating mother should in her refusal to sustain life (Watkins 2015, 7), and in evoking this metaphor, Watkins asserts that "Mother Nature" owes us nothing. *Gold Fame Citrus* refuses any assurances of wilderness as refuge and not least because its wilderness prophet is exposed as a "fraud" and huckster, stealing water and supplies while pretending that he has been dousing water through his spiritual connection to nature (286–287). Luz allows herself to believe he is a man with a radically transformative vision, yet once she stops taking Levi's hallucinogens, she realizes he is just another self-interested survivalist who deploys threats, coercion, and sexual control of his drugged harem of women to retain power over his followers who have nowhere else to turn. His wilderness utopia is "a madman's colony, an outpost in the cruel tradition of outposts, peopled by prostitutes and loners and rejects and criminals and liars, their sheriff a con and a thief and surely worse" (312). The frontier myth, Watkins points out, has only ever been a myth of American settler supremacy founded on the reality of violence, theft, and cruelty.

The New Wilderness and *Gold Fame Citrus* deflate any hope that the myth of wilderness will turn out to be a real place of refuge and safety where human survival skills—or pioneer ingenuity and individualism—triumph over the hardships of nature. Watkins and Cook refuse to romanticize human survival skills as a value in and of itself through characters who mostly endure the postapocalypse due to random chance. For example, Luz has no innate survival skills that would protect herself or Ig, and she constantly relies on others to take care of them. In *The New Wilderness*, Community members die from a variety of natural hazards—falling off cliffs, drowning in a river, eaten by a cougar—leading Bea to realize that "they'd escaped one monster by hiding in a closet, only to find another there among the hangers, claws unsheathed" (Cook 2020, 20). Wilderness or nature, in other words, is not a benevolent mother invested in human survival but an inhuman, overwhelming force, raw and red in tooth and claw. Like Luz, Bea is disillusioned of the "cultural belief" that close contact with nature inherently makes anyone safer or "better" and wonders perhaps if better "might have only meant better at surviving, anywhere, by any means" (36). Ultimately, the maternal ambivalence of Bea and Luz, who abandon their roles as the "good" mother when they abandon their daughters and respective cults of wilderness, indicates a rejection of those myths that are incapable of imagining human survival beyond patriarchal settler terms of conquest, competition, and control.

Nomadic mothers and posthuman wilderness ethics

Gold Fame Citrus and *The New Wilderness* do not merely critique cultural myths of wilderness and motherhood but attempt to offer a posthuman vision of a wilderness ethic that situates human survival in *relation* rather than *opposition* to nonhuman survival. In their engagements with the history of American environmental and wilderness preservation movements, the novels reject Muir's wilderness ideal of a "pure" nature in terms of "mother-love" in favor of Aldo Leopold's more grounded "land ethic," outlined in his seminal work, *A Sand County Almanac* (1949). Leopold's central "ecological insight" about wilderness preservation "suggested the need for a new approach, based on ethics, that would make men aware that their environment was a community to which they belonged, not a commodity that they possessed" (Nash [1967] 2014, 192). Although Leopold's insight was not necessarily "new" in its alignment with Indigenous and First Nations traditions and land management practices, it was a radical departure from settler views of conquering the wilderness for economic profit, a utilitarian philosophy that dominated much of the conservation movement. An employee of the National Forest Service, Leopold spent years working within and struggling against those same paradigms until his view of "wilderness preservation broadened and deepened into a philosophy of [human] responsibility to the rest of life" (Nash [1967] 2014, 191–192). We might read Leopold's shift in perspective as illustrative of contemporary ecofeminist and posthumanist calls for shifting our views of human-nonhuman relations. As Stacy Alaimo (2016) contends, and reminiscent of Leopold's insights, if we accept humans are "immersed and enmeshed in the world" (157), this generates "a sense of species in relation" that makes possible a posthuman wilderness ethic (151).

Drawing from Leopold's land ethic and extending this to a posthuman wilderness ethic provides insights into how Watkins and Cook explore the limitations of wilderness and motherhood mythologies *and* suggest points of resistance that might help dismantle ways of being in the world that are no longer sustainable when confronting the realities of the Anthropocene. Faced with climate collapse, resource scarcity, and environmental precarity, Bea and Luz realize no matter what choices they make, there is no benevolent nature to return to, and there is no mode of mothering available to them within the essentialized and constructed myths of motherhood that they thought might sustain them and their children in the wilderness. They are confronted with the actual reality of wilderness spaces where survival is a constant challenge and never guaranteed and where they must accept the continuation of life according to very different posthuman terms that emphasize human vulnerability and interdependence with nature.

The New Wilderness immediately opens with the impossibility of survival, as Bea delivers a stillborn child and fends off animals from feeding on its corpse. Even if giving birth is simply part of "the work of survival," the sadness and relief Bea feels over her failure to bring another child into "this world" signals the brutal conditions of wilderness life and Bea's eventual choice to abandon her living daughter, Agnes (Cook 2020, 4, 5). Although she loves her daughter "fiercely," she has felt conflicted by the burden of motherhood since Agnes's birth, viewing it as a "heavy coat" she is "compelled" to wear (18). When Agnes becomes sick from the City's pollution, prompting their move to the Wilderness State, Bea resents the implicit expectation that she must sacrifice her own comfort and safety for the sake of her daughter's health: "*So, we have to risk all our lives just to save hers? Is this the rule, or do I have a choice?*" (72, italics in original). As much as she tries, Bea cannot "let go of her own self […] to hold up Agnes" (73). She struggles to maintain a sense of maternal responsibility, at least within the terms of the world she and Agnes have inherited, where "new life" is impossible because there are too many people trying to use too few resources (122). Thus, what drives Bea's choices and maternal ambivalence is a profoundly changed world, at least one in which the relatively privileged experience of traditional white motherhood is no longer available to her.

Bea cannot be the "good" mother she imagined she would be, and so she attempts to be a "wilderness mother" who prioritizes the survival of her offspring but only until the child can survive independently and "find their own land to explore" (Cook 2020, 281). This wilderness model of motherhood, where a mother leaves her child without regret, is no more convincing than the all-sacrificing myth of the "good" mother. Bea cannot stop longing for her own mother, left behind in the City, and when she learns her mother has died, she abruptly abandons six-year-old Agnes to fend for herself as a member of the Community. Now motherless, it appears Bea no longer felt obligated to remain a mother to Agnes, "who seemed not to know what love was, who had turned too wild to know it" (132–133). Agnes attempts to rationalize Bea's abandonment by convincing herself that motherhood should be animalistic and driven by instincts of self-preservation, but when Bea eventually comes back, Agnes cannot understand her sudden absence or return: "Why did her mother insist on being so many people at once when Agnes only needed her to be the one?" (338). Agnes expresses here the desire for Bea to fulfill the myth of the "good" mother: a constant, self-sacrificing, loving source of protection and plenitude. Bea's failure to be the "good" mother is also linked to the failure of the Wilderness State, and "mother earth," to fulfill its mythical promise of safety and sustenance. Just as the wilderness is indifferent to human needs and survival, Bea is indifferent to her daughter's needs and rejects motherhood in a seemingly unforgiveable act of selfishness and cruelty.

However, we might read Bea's maternal ambivalence and rejection of motherhood more positively through the lens of "nomadic subjectivity," which Rosi Braidotti (2010) argues is a mode or process of "becoming other" in posthuman terms (208). For Braidotti, posthumanism signifies "other" ways of being in the world that enact a decentering of "the human" through a rejection of anthropocentricism and human exceptionalism, the two dominant perspectives or modes of thinking that many critical ecofeminists argue to be responsible for the ongoing environmental catastrophes of the Anthropocene. To be posthuman, then, is "to experiment with different modes of constituting subjectivity and different ways of inhabiting our corporeality" by accepting human/nonhuman "interrelationality" (209). The process of becoming nomadic/other in posthuman terms is therefore an affirmation of the complexity of life beyond binary categories and from a position of "vulnerability" and "multiplicity" when confronted with "the concept of limit" (209–210). In other words, according to Braidotti's posthuman terms of becoming nomadic, Bea comes up against the painful limits of her subjectivity, as a woman and mother, and reaches the point where one "can think / understand / do / become no more than what he or she can take or sustain within his or her embodied, spatiotemporal coordinates" (210). When Bea abandons Agnes, she is no longer capable of enduring the conditions of bare life in the Wilderness State, where *zoē*, or "life as absolute vitality" (210), is a "generative process" that also incorporates death and strips the subject of any illusions about human survival as the central concern of "Life" (212). Bea sees no possible alternative for mothering within these contexts, or at least no possibility of motherhood that fulfills the demands of ensuring the survival of her child above all things when the world they inhabit has reached the brink of extinction because of a settler colonial history of prioritizing human survival over any other forms of life.

Indeed, Leopold warned that "disharmony and sickness would continue to characterize those parts of the earth man had civilized" unless there was a change in "man's" view of nature, from one of conquest and consumption to acceptance of the "true place" of humans as dependent upon and part of "the biotic community" (Nash [1967] 2014, 196). The central tenets of Leopold's land ethic are akin to Braidotti's (2011) argument for a nomadic/posthuman ethics that resists anthropocentricism by grounding "the subject in a materially embedded sense of responsibility and ethical accountability for the environments s/he inhabits. What is at stake is the very possibility of the future, of duration or continuity" (112). Bea sees what is truly at stake in the experiment of the Wilderness State: not the priority of human survival but the need to create the possibility of a sustainable future for the life of the planet. Those stakes are even more urgent when confronting ongoing ecological challenges and the enormous scale of species and habitat loss that define the realities of the Anthropocene. Thus, as a crucial response to the Anthropocene, becoming nomadic/other in Braidotti's posthuman terms is an ethical choice invested in a livable future while grounded in "a livable present" (237) that accounts for "the shared capacity of humans to feel empathy for, develop affinity with, and hence enter into relation with other forces, entities, beings, waves of intensity" (317).

The eco-ethical philosophies proposed by Leopold and Braidotti insist on a moral responsibility to the world and our specific environments that requires resituating human relationships with nonhuman nature; instead of assuming human superiority and separation from nature, which has led to dominance, exploitation, and environmental destruction, we need to engage in vital practices of interconnection, mutualism, and care. This might have been the real potential of the Wilderness State. Rather than a mythic place of refuge, it is an actual space where human interaction with nature, or life as absolute vitality, is grounded in the difficult but necessary relations of mutualism and interdependence. The irony, of course, is that the primitivist cult of wilderness in *The New Wilderness* cannot survive without depending on its interconnection with nonhuman nature, yet instead of forming ethical relations with the land that might translate into more ethical relations between humans, the Community devolves into cruelty, competition, and abject despair over its dependence on nature to survive. They cannot "see the land as a whole," or themselves as part of the whole, which Leopold argued was central to his land ethic—our ability "to think in terms of community rather than group welfare" and to engage with the land beyond the concerns of short-term survival and benefit (cited in Nash [1967] 2014, 196). Although Bea's decision to abandon Agnes appears cruel and selfish, driven by her own sense of short-term welfare, it presents a moment of radical insight into how the Community's literal nomadism is unsustainable; it offers nothing different from a ruthless survivalism predicated on human domination, self-interest, and disconnection from nature.

Bea's maternal abandonment is as an act of resistance that refuses the terms of mythological motherhood or being the "good" mother; she acknowledges the necessity of "becoming other" or "becoming-woman" in nomadic terms, which Braidotti (2011) insists subversively work "toward the transformation of [...] the embodied histories of white institutionalized femininity" (49). Bea's choice to abandon Agnes is a deeply ethical one that frees her and Agnes from any loyalty or "bondage" to the dominant ideologies of mythological (white) motherhood, especially when the threat and fear of extinction present challenges in thinking "beyond survival and mortality" or imagining possible alternatives or futures (330). As Braidotti reminds us, a nomadic ethics of becoming-other "is a mode of actualizing sustainable forms of transformation," and although it may not relieve our fears of extinction, the pain of accepting our "vulnerability" requires us "to think about the actual material conditions of being interconnected and thus being in the world" (316–317). Bea must leave so that she can become nomadic/other, refusing the binary choices and myths of the "good" or "bad" mother so that she might shift her perspective and become a different, perhaps "better" mother. Admittedly, it is easier to read Bea's maternal abandonment as an affirmative rather than negative choice because she returns to Agnes, transformed as a nomadic mother who "left" but "came back" and "loved" her daughter who has no obligation to "forgive" her mother (Cook 2020, 368). Bea is simply present but not endlessly available or reassuring; instead, she offers Agnes the space to define her own terms of identity but without any redemptive or utopian hope invested in motherhood or the future.

Certainly, the ending of the novel appears quite bleak, with Agnes and her adopted daughter, Fern, removed from the Wilderness State—now turned into Private Lands—and deported back to the City. They are unlikely to survive, but Bea's earlier revelation when Agnes confronts her for leaving—angry, wounded, yet admitting her enduring love for her mother—is the locus of hope in the novel: "'What a wonder you are,' she heard her mother say, not to her, but to the air, the lands below, the sky, the forest, to herself. [...] 'Look at the *wonder*. I *was* a good mother'" (Cook 2020, 369, italics in original). Bea became a good, as in ethical, mother because she gave them both the freedom to be wild, or bewildered, by the pain, vulnerability, and joy of being-in-the-world—"the air, the lands below, the sky, the forest"—subject only to "the politics of life itself as a relentlessly generative force" (Braidotti 2011, 333). By the time Agnes leaves the Wilderness, which is no longer a wilderness, she understands, "All natural things are known and understood somewhere inside a natural being" (Cook 2020, 389), and that even the City "is just another wilderness. [...] A thing to become part of" (392). She also "began to know [her] mother" (394), as she recognizes her own maternal ambivalence—the fear, disgust, and overwhelming love for her child—and accepts these "complications and confusions because those complications and confusions are what makes them true" (395). Agnes and Bea accept the reality rather than myth of motherhood; maternal love is complicated, confusing, and never a guarantee of ensuring the survival of one's child; mothering is an ongoing process of becoming and always in the contexts of shifting environments and circumstances that may be hostile to life (*bios*) but are nevertheless part of the persistent regeneration of Life (*zoē*) beyond human concerns.

Luz in *Gold Fame Citrus* might also be read as a nomadic mother but in more challenging terms than Bea. Unlike *The New Wilderness*, in which Bea and Agnes redefine for themselves the expectations of motherhood and their relations to nature, *Gold Fame Citrus* offers no viable alternative to mythologies of mothers and "mother earth." Instead, the novel charts the collapse of these ideals and leaves the reader with the conclusion that survival, let alone thriving, is impossible in the context of ecological catastrophe and that motherhood is made untenable by this collapse; we are headed for extinction, there is no point in resisting this inevitability, and thus engaging in wilderness and motherhood is equally futile. However, Braidotti (2011) insists on the need to adopt "a mode of thinking that [...] aims to think with and not against death," while emphasizing "the importance of maintaining a refusal that turns into an affirmation [...] that undoes existing arrangements" (335). Watkins, and by extension Luz, refuses all "existing arrangements" as long as our constructs of motherhood and wilderness remain mired in negative fantasies or histories of patriarchal settler relations to the environment, where wilderness and mothers are imagined as endlessly available yet unreplenished sources of life. To become truly nomadic, Braidotti (2011) argues, requires a confrontation with death, though not as an end to life but as an affirmation of continual processes of regeneration, wherein humans are part of "a vital web of complex interrelations," simply one species "among multiple others in the flow of monstrous energy of a 'Life' that does not respond to our needs" (53). Luz may seem to be a monstrous mother but only insofar as Life, and wilderness, and nature, and mothers are perceived as monstrous when they fail to live up to human expectations, desires, or myths.

Luz appears to be an "unnatural" mother from the start of the novel, when she impulsively kidnaps Ig, a somewhat feral toddler, because she feels "an unbearable welling of affection," though there is no logical or "natural" reason for this connection (Watkins 2015, 31). For a while, the ideals of motherhood carry Luz and Ig (and Ray) through several weeks of "blissed-out chaos up in the canyon. [...] But it could not last. (Nothing here could.)" (54). The California climate, its environment, and the world at large have changed to the degree that "nothing" is sustainable. It is the acknowledgement of this reality—the impossibility of survival without "sustainable processes of transformation" (Braidotti 2011, 231)—that guides Luz's choices. Luz is not a "bad" mother so much as a mother who is unable to believe anything is capable of enduring: "One didn't cultivate faith and one did not cultivate anything here" (Watkins 2015, 94). Luz thinks that if she could sustain faith in a future, she could be "better" and provide Ig with a better life. Motherhood is not transformative in this novel but an illusion, an act of faith, a role that Luz pretends she might fulfill if she just believes hard enough. Once she and Ig arrive at the Amargosa colony, Luz tries to buy in to the patriarchal myth of motherhood and plays at being a "good mother," which merges into being both "pure" and a "good woman" (168). Luz knows she cannot be any of those things, though, and not because this ideal of a pure/good mother is impossible but because being a good or bad mother makes no difference in changing the circumstances or outcome of their situation. The climate catastrophe was already well under way before Luz was born and the possibility of a better sustainable future a bankrupt endeavor.

Luz also fails to believe in the possibility of being the "good" mother partly because she failed at being the good child, reified since birth as the symbolic promise of ecological renewal and fertility. Luz was "adopted" by California's Bureau of Conservation as the "poster child" for all its "promises vague and anyway broken" when it announced upon her birth "*an heroic undertaking which will expand the California Aqueduct a hundredfold, so that Baby Dunn and all the children born this day and ever after will inherit a future more secure, more prosperous, and more fertile than our own*" (Watkins 2015, 10–11, italics in original). Watkins' language serves as a direct critique of the preservationist movement and the 1964 Wilderness Act in their calls for saving wilderness for future generations. Luz's life has been defined by her role as symbol of a larger, and ultimately failed, project to transform California, and the American West, from a sterile desert to a fertile garden. Baby Dunn is meant to give hope to an entire generation of people and make them believe in the impossible—that the environment can always be shaped to human wishes and desires with the use of just a little more force or technology. Like the land she is meant to represent, though, the fantasies projected onto her as Baby Dunn leave Luz in ruins, the "goddesshead of a land whose rape was in full swing before she was even born" (12). Without her consent, Luz becomes conflated with the land she is meant to represent.

It is no wonder, then, that Luz resists Levi's environmental (eco-terrorist) propaganda campaign, which would cast Ig as a new Baby Dunn and Luz in a new symbolic role: "We need you both, and we need it to be big and wholesome and beautiful. Transcendent. Madonna and child" (Watkins 2015, 229). Here, Levi asserts two equally powerful mythologies in American culture: wilderness and motherhood. Within this framework, Luz would function as the (sullied) symbol of a corrupt and failed environmental project who also has been transformed into a mother, thereby both redeemed and redemptive. Ig, and by extension the world Levi hopes to create without any obvious strategy but violence and theft, will represent a "fresh start." Luz instinctively knows, "urgently, desperately, painfully," that Levi's proposal will be harmful and useless, exploiting Ig as an empty symbol of futurity (231). Levi's vision is just as flawed as the American wilderness ideals that Luz cannot believe in, knowing that equating maternity with a redemptive philosophy of the natural world is dangerous at the individual and collective level.

However, when Levi insists on keeping Ig as his own poster child, Luz must choose between leaving with Ray or staying with the commune. Luz chooses to abandon Ig and immediately feels "the astonishing relief of quitting. Taking her rightful position in that long line of runners and flakes. Those were her people. [...] Antsy pioneers, con artists and [...] dowsers and gurus" (Watkins 2015, 337). Luz accepts that she is part of the larger collective of those who turned the frontier into the corrupt mythology of the wilderness, and she no longer needs to be something different or pretend that survival under these terms is possible. When she and Ray are caught in a flash flood, she succumbs to the rushing water and, before she is swept under, insists that she would "be okay [...] if I could just get my feet under me" (339). Luz, in other words, would choose to survive if only the world were otherwise. Her death, though, can be read as an affirmative choice to accept herself as a nomadic subject, part of the monstrous flow of Life with its relentless aim "at self-perpetuation and then, after it has achieved its aim, at dissolution" (Braidotti 2011, 348). The "miraculous" appearance of water in the desert, and Luz's "serene" yet insistent desire to become part of its "roar" and "rage"—as the "flood came upon them like an animal, like a vengeful live thing, earth-colored and savagely fast"—is the only redemptive moment for Luz (Watkins 2015, 338–339). Her choice embodies the "paradox" of a posthuman ethics of accepting death-in-life and life-in-death, wherein the human "struggle for survival" gives way "to surrender[ing] the self [...] thus choosing our own way of disappearing, our way of dying to and as our self" (Braidotti 2011, 349). If Luz was trapped and reified since birth by the mythological symbol of land-as-woman, her willful death transforms her into a living participant, however briefly, in the flooding rush of Life. She accepts her true self as only an instance of desire and duration in the unrelenting processes of the Earth's self-perpetuation and dissolution.

Troubling mythologies: extinction, ecological grief, and restoration

Gold Fame Citrus and *The New Wilderness* resist reassuring or mythical possibilities for motherhood and wilderness survival when confronted with environmental catastrophe. They prompt an interrogation of the primary assumption that undergirds mythologies of motherhood and wilderness: that both will allow humans to survive, and even thrive, in a postapocalyptic world of ecological collapse. As explored throughout this chapter, myths of wilderness and motherhood are no longer available as sites of refuge, nor have they ever been because in reality these myths are deeply implicated within patriarchal and settler colonial histories of violence, exclusion, and oppression. Moreover, these mythologies undergo a deep shift when placed within narratives of impending extinction, where the resources needed for survival are no longer available in a world that is changing in devastating ways. Bea and Luz try to fulfill the role of the "good" mother, who will protect, nurture, and sacrifice herself for her children; both believe they are saving their children by bringing them to wilderness spaces, tentatively buying into the myth that the wilderness will provide a haven from the effects of urban decline and the climate crisis. They discover there is no escaping the extent of ecological loss and harm and that the ideals implicit in myths of wilderness and motherhood are no longer tenable, if they ever were, because the inscription of a feminized or maternal nature throughout America's history of settler colonialism has only led "to psychological patterns of response that cannot possibly result in anything other than abuse" (Kolodny 1973, 180). Bea and Luz realize that relying on myths of wilderness and motherhood without changing those patterns has only brought their world to the brink of extinction, making the terms of survival and motherhood all but impossible.

In this sense, the survivalist postapocalyptic settings of the novels implicate the Anthropocene as the inevitable endpoint of a settler colonial history. The conquest of the American frontier, responsible for the depletion of natural resources, disappearance of wilderness spaces, and devastation of complex ecologies that make possible healthy, biodiverse environments, leads to the collapse of civilization and the hope that wilderness/mother earth will provide the possibility of a new start. However, just as actual mothers rarely live up to the myth of motherhood as infinite abundance, actual wild spaces rarely offer inhabitants what they are hoping to find. For Bea and Luz, the ability to raise their daughters and hope for a better life, or even a life that is not impossible, diminishes over the course of the novels. Unable to mother their daughters under such intolerable conditions—or at least be the mothers they are expected to be—leads them to abandon their children abruptly and without explanation. Their maternal abandonment seems to be a natural response of ecological despair and grief, a profound loss of faith in the possibility of survival. However, the narrative implications of maternal ambivalence and abandonment in both novels prompt the demand for new mythologies that imagine how we might survive in a world wherein neither mothers nor our environments provide us with what we need.

Through the characters of Bea and Luz, the novels engage with the complexity of ecological mourning and grief as both a debilitating experience and yet a potentially radical or subversive source for transforming ethical relationships, subjectivities, and acts of resistance in the face of extinction. According to Ashlee Consolo Willox (2012), mourning the loss of habitats, species, and places due to the climate crisis is a powerful motivator for changing our views of how we are situated in relationship to nonhuman nature. Likewise, Ursula Heise (2016) argues that environmentalist movements and narratives need to be affirmatively future-oriented and provide "visions that are neither returns to an imagined pastoral past nor nightmares of future devastation" (12). We require instead a postanthropocentric, posthuman ecological conscience and ethic that goes beyond mourning loss to broaden our "understanding of extinction not only as a narrative endpoint, but as a possibility of new beginnings—not the end of nature so much as its continually changing futures" (54). Watkins and Cook provide timely extinction narratives in response to the Anthropocene that suggest the necessity of accepting that the dominant myths of motherhood and wilderness are no longer viable. There is no all-encompassing, all-loving, and all-sacrificing mother, nor is there an abundantly life-sustaining wilderness that provides us with refuge from the catastrophic mess a history of settler colonialism has made of our environment.

This is not to imply that we give up on wilderness; rather, the urgencies and escalating environmental crises brought on by the Anthropocene demand that we cultivate a posthuman wilderness ethic. Bea's and Luz's choices to abandon the cult of wilderness and the mythical ideal of the "good" mother in favor of becoming nomadic mothers illustrates a necessary ethical shift when telling narratives about human survival. We need to relinquish human-centered wilderness survival myths and create eco-centered stories through a posthuman framework "that embraces the wild, even as it is wary of wilderness paradigms that divide humans from nature" (Alaimo 2008, 252). In other words, as an alternative to settler colonial and patriarchal myths about returning to wilderness as refuge, we might emphasize ongoing vital practices of restoration and rewilding. That is, restoration as an ethical act of mitigating centuries of harm done to wild places and their inhabitants—thus, acts of restorative justice—and restoration not purely in the interests of human survival but because "the survival of many species depends on creating more areas in which wild creatures and ecosystems can flourish" (258). If we look back to Leopold's land ethic, he provides a similar argument in his insistence on the necessity of wild places, which reveal the continuity of land through time and our "actual relation to the natural world" as one of mutual survival and care (Nash [1967] 2014, 198–199). It is contact with wilderness that instills within us an ethics—not the myth of wilderness that entails a pure, pristine, untouched nature beyond human impact but a wilderness that involves human participation in the biotic community. Employing this wilderness ethic strengthens our understanding of humans as part of the whole ecology, where human survival and the ability to thrive is reliant on an ecological engagement aimed at responsible restoration, recognition of interdependence, and leaving things in a better condition than we found them. This seems an equally productive model for mothering in a world marked by environmental precarity yet where we constructively engage in creating a livable present that might make possible a livable future.

Notes

1 Influential studies of feminist theory critiquing maternal myths include Nancy Chodorow (1978), *The Reproduction of Mothering: Psychoanalysis and the Sociology of Gender*, Berkeley: University of California Press; Sarah Hrdy (2000), *Mother Nature: Maternal Instincts and How They Shape the Human Species*, London: Vintage; Adrienne Rich (1976), *Of Women Born: Motherhood as Experience and Institution*, New York: Norton; and Sara Ruddick (1989), *Maternal Thinking: Toward a Politics of Peace*, New York: Ballantine Books.

2 Although a more detailed discussion is beyond the scope of this chapter, it is important to note "racial and class biases in the social construction of good and bad mothers" and that the experiences and myths of traditional motherhood, "as defined by dominant motherhood ideologies," rarely apply to poor, working-class women and members of BIPOC and LGBTQ communities (Johnston and Swanson 2003, 22).

References

Alaimo, Stacy. 2008. "Trans-Corporeal Feminisms and the Ethical Space of Nature." In *Material Feminisms*, edited by Stacy Alaimo and Susan Hekman, 237–264. Bloomington and Indianapolis: Indiana University Press.

Alaimo, Stacy. 2016. *Exposed: Environmental Politics and Pleasures in Posthuman Times.* Minneapolis: University of Minnesota Press.

Baker, Jan. 2013. "What Have We Done to Mother Earth? Psychodynamic Thinking Applied to Our Current World Crisis." *Psychodynamic Practice* 19, no. 1: 55–67.

Braidotti, Rosi. 2010. "The Politics of 'Life Itself' and New Ways of Dying." In *New Materialisms: Ontology, Agency, and Politics*, edited by Diana Coole and Samantha Frost, 201–218. Durham and London: Duke University Press.

Braidotti, Rosi. 2011. *Nomadic Theory: The Portable Rosi Braidotti.* New York: Columbia University Press.

Comer, Krista. 1997. "Sidestepping Environmental Justice: 'Natural' Landscapes and the Wilderness Plot." *Frontiers: A Journal of Women Studies* 18, no. 2: 73–101.

Cook, Diane. 2015. *Man v. Nature: Stories.* New York: Harper Collins.

Cook, Diane. 2020. *The New Wilderness.* New York: Harper Collins.

Crane, Kylie. 2012. *Myths of Wilderness in Contemporary Narratives.* New York: Palgrave Macmillan.

Finney, Carolyn. 2014. *Black Faces, White Spaces: Reimagining the Relationship of African Americans to the Great Outdoors.* Chapel Hill: University of North Carolina Press.

Heise, Ursula. 2016. *Imagining Extinction: The Cultural Meanings of Endangered Species.* Chicago and London: The University of Chicago Press.

Johnston, Dierdre D., and Debra H. Swanson. 2003. "Invisible Mothers: A Content Analysis of Motherhood Ideologies and Myths in Magazines." *Sex Roles* 49, no. 1/2 (July): 21–33.

Kaminski, Johannes. 2021. "The Neo-frontier in Contemporary Preparedness Novels." *Journal of American Studies* 55, no. 4 (June): 1–29.
Kolodny, Annette. 1973. "The Land-as-Woman: Literary Convention and Latent Psychological Content." *Women's Studies* 1: 167–182.
Leopold, Aldo. 1949. *A Sand County Almanac: And Sketches Here and There.* Oxford and New York: Oxford University Press.
Nash, Roderick Frazier. (1967) 2014. *Wilderness and the American Mind.* 5th ed. New Haven and London: Yale University Press.
Sutter, Paul S. 2021. "Putting the Intellectual Back in Environmental History." *Modern Intellectual History* 18: 596–605.
Theriault, Daniel, and Rasul A. Mowatt. 2020. "Both Sides Now: Transgression and Oppression in African Americans' Historical Relationships with Nature." *Leisure Sciences* 42, no. 1: 15–31.
Watkins, Claire Vaye. 2012. *Battleborn.* New York: Riverhead Books.
Watkins, Claire Vaye. 2015. *Gold Fame Citrus.* New York: Riverhead Books.
Watkins, Claire Vaye. 2021. *I Love You but I've Chosen Darkness.* New York: Riverhead Books.
Willox, Ashlee Consolo. 2012. "Climate Change as the Work of Mourning." *Ethics & the Environment* 17, no. 2: 137–164.
Wohl, Ellen. 2013. "Wilderness Is Dead: Whither Critical Zone Studies and Geomorphology in the Anthropocene?" *Anthropocene* 2: 4–15.
Zimmerman, Michael E. 1995. "The Threat of Ecofascism." *Social Theory & Practice* 21, 2 (Summer): 207–238.

PART II

Indigenous and Afro-diasporic myths and ecological knowledge

3

BOUNDLESS WATER, BOUNDLESS ICE—ARCTIC COSMOLOGICAL CONCEPTS IN TIMES OF MELTING HORIZONS

Sonja Ross

A shift of the Earth's axis?

The Fridays for Future movement has drawn great attention to the phenomenon of global warming within the last four years. During the accompanying intense and ongoing public discussions, a strange message suddenly came to my mind. It was a short headline that had caught my attention a few years earlier: Inuit elders would, as I recalled, attribute global warming to a shift in the Earth's axis and thus not to human-induced CO_2 emissions. They said the Earth's axis must have shifted because in former times it used to have approximately one hour of daylight for seal hunting but today it is two hours. The Earth wobbled and therefore the sun, moon, and stars are out of place. Moreover, there is more southerly wind than northerly wind, which would cause serious changes in the structure and appearance of the environment. This interpretation was surprising and disturbing at the same time because at first glance it meant that a geo-physical process would cause global warming, not human influence. The Inuit message of the tilted axis has been circulated and cited in numerous ways. Finally NASA got involved and offered their view.

Data from long-term studies do indeed suggest that there was an extraordinary pole shift to the east after 1990, as we will see later, although this movement can only be explained by a global human influence on the climate and atmosphere. Nevertheless, the Inuit perspective was a matter of "thinking big," and obviously it came from an understanding of larger contexts. It was based on knowledge and experience given how precise the justifications were made based on astronomical observations and concrete weather phenomena. This example shows that coherent results can be obtained from different perspectives. Objective evidence, however, also requires robust data and comparative analysis. So,

DOI: 10.4324/9781003348535-6

how can the constitutive elements of this narrative, which is based on factual observations, be understood?

In their hunting way of life, the Inuit are directly dependent on a seemingly boundless landscape shaped by wind, snow and pack ice, seasonal stability, and the predictable availability of birds and game, marine mammals, and fish stocks. In their cultural memory, they could rely on environmental and climatic conditions as the basis of their hunting habits, their movement patterns, their festive annual cycle, and their social activities. The cultural substructure seems to have remained stable in the traditions and conventions across generations. The values and knowledge of most elders who were born on the land and are still alive were shaped by largely predictable changes in the environment (ITK 2019, 10). Over several centuries, the great waves of change brought about by the European (including Russian) and American fur trade, the whaling industry, the nationalization of arctic territories and re-education of the local population, as well as the current promise of inexhaustible energy sources for gas and oil exploitation in the more northerly polar regions, brought about rather selective adjustments in lifestyle.

> Indigenous peoples (have) a rich and extensive repertoire of stories, mythologies, and beliefs pertaining to the Arctic and beyond—about humans, animals, land, water, ice, air, wind, the sky and the atmosphere. These stories matter because they give shape to those communities and help guide current and future relationships with northern settlers and national governments.
>
> *(Dodds and Nuttall 2019, 86)*

Narrative traditions and the mythological superstructure, however, adapt almost imperceptibly to the reality of life (Ross 2000). Processes of change can only be traced with ethno-historical or cultural–historical methods. As a rule, we do not know what has already been lost in the remembered time and where exactly the reference point for a specific transformation analysis should be set. Moreover, the demand to reproduce narratives mnemonically only in exact wording and with exact gestures (Hall [1939] 1975, 412–413) requires special empathy and linguistic competence on the part of field researchers. Especially in times of colonial re-education, which forced a condensed cultural transformation, the tracing of omitted fragments or episodes was necessary. Rasmussen ([1932] 1949, 7), using examples from Alaska, lamented that one must listen very carefully to detect the disappearing sounds within a colonial setting: "[…] the old hunters have now become reindeer herders and the lore of pagan times was therefore not very present." It takes a lot of patience to attract old storytellers with highly developed performing skills. The influence of the surrounding Indians and the Chukchi of Siberia can be felt, but the "peculiar Eskimo undertone," which we know so well from the myths and sagas of Greenland, runs through all the narratives.[1] So it is by no means certain that everything that seems to be autochthonous is actually indeed autochthonous.

Changing planetary and stellar conditions such as the position of the Earth in relation to the sun—this was the Inuit's message—led to a warming of the atmosphere and thus to massive effects on the environment and fauna. For evidence-based analysis, it is important to determine, if possible, the original occurrence of the information "Earth axis" in its context. As will be seen, the Earth's axis has a presence in internet media that far exceeds its original significance as an etiological reference in the region. As soon as sufficient information can be gathered in its context, it becomes possible to understand individual strands of information in their empirical order, their possible derivations, and their meaning in the narrative context.

Through a whole series of repetitions, derivations, copies, and secondary information,[2] my investigations finally led me to focus on an initiative from the years 2000–2001 (Dickie and Woolf 2000; Ashford and Castleden 2001), in which Western scientists together with local Inuit in northern Canada conducted a study on climatically induced processes of change. In the further development and intensive involvement of the population with the methods thus gained—and climate conditions changing very rapidly at the same time—in 2010, independent of the first study, another documentary film was produced in which the Earth axis theory was finally named and foregrounded (Mauro and Kunuk 2010).

However, at the time, more than 30 years ago, the international community had already begun to establish a well-founded definition of the causes of climate change and to develop countermeasures.[3] A theory of the movement of the Earth's axis, which would cause longer solar radiation in the polar region, seemed to therefore have little relevance in the scientific discussions in view of growing data collections and overwhelming evidence that the immense CO_2 emissions of industrial nations cause the rapid climate changes. Nor is there any evidence of the idea and use of the Earth's axis as a figurative element in traditional arctic worldviews or in the ethnographically described ecstasy techniques and extraterrestrial flights of shamans—comparable, for example, to the motif of a sky ladder that was widespread in the Eurasian region. And yet, in explaining the drastic changes in climate and environmental conditions, adaptation and application of this element follow a coherent concept. It uses traditional star lore as well as the evolved practical knowledge of environmental and living conditions and then combines traditional conventions with Western scientific techniques.

Environmental issues

The polar region of the Arctic is the area of the world that was most affected by global warming. As Rosemarie Kuptana,[4] the former head of the ITK,[5] the national organization for 65,000 Inuit living in Canada, asked in the documentary *Sila Alangotok: Inuit Observations on Climate Change*[6] (Dickie and Woolf 2000): "Are we the messengers for the rest of the world?" Because of the worrying effects of global warming on the lifestyles of local people, the IISD (International Institute for Sustainable Development) conducted studies in cooperation with 16 households on Banks Island from 1999 to 2001 to raise public awareness of the urgency of the issue and to link traditional knowledge, local adaptation strategies, and scientific analysis (Ashford and Castleden 2001). The team of natural scientists and representatives of governmental and wildlife organizations as well as a film team was complemented by 45 community knowledge experts, who together collected observations on the changing environmental conditions and analyzed their impact on local lifestyles. About 20 years ago, the observations were already alarming, and the interviews were marked by great uncertainty about the future. The environment and its landscape characteristics are no longer familiar, the weather is unpredictable, the seasonal cycles are shifted, summers are generally warmer and longer, winters are no longer cold enough to maintain the multiyear ice, the sea ice is frozen for only a short time, and the coastlines are fragile because of the melting permafrost, which thus also endangers places and infrastructures. Today, all of this sounds oppressively familiar to us. But for communities that link their economic subsistence, their knowledge, their living conditions, and their value system with their environment and its resources, this is vital. In the results of the project, the researchers make it clear that the Inuit have extensive knowledge of the "historical and current landforms, erosion activity and permafrost conditions" and that these are "closely tied to community activities, including travel, hunting and fishing" and cannot be considered in isolation from other "variables like wind, precipitation, temperature, human activity and seasonal change" (Ashford and Castleden 2001, 16). Not surprising, further changes concern the composition and quantity of huntable animals on water, on land, and in the air as well as the increased occurrence of insects and fundamental problems in finding and bringing in prey. With the public attention paid to this study and the accompanying scientific discussion, the importance of traditional knowledge (TK) and traditional ecological knowledge (TEK)[7] in the indigenous perspective has increased and consolidated. Inuit communities made organized efforts to systematize the experiences from cooperation with the scientists and to develop (ITK 2019) their own adaptation strategies. In addition, the collected and structured knowledge was to be conveyed in training and further education. In doing so, a comparison with "Western" natural sciences was deliberately made, and an intersection of "common ground" of categories that apply to both sides was formed (Sutherland 2003, 14).

The differences identified are remarkable: On the side of "Western" sciences we can find discipline-based micro and macro theories and mathematical models with evidence in the physical world as well as quantitative, tool-based written reports. In the juxtaposition, TEK appears with qualitative oral traditions, holistic reflections on physical and metaphysical existence, trust in the inherent wisdom of the traditions, their rootedness in daily life, and respect for all things. These lists of "epistemological contrasts" show precisely the weak point of studies on traditional environmental knowledge, namely, their restriction to questions of "scientifically encodable forms of knowledge and their relation to subsistence," as Koester stated in 2002. Thus, in his discussion of the studies of TEK in the Kamchatka region, he raised the most relevant question about "Traditional Environmental Knowledge versus Knowing and Living Traditionally Environmentally" (Koester 2002, 48–49).

Despite emerging concerns, even in the 2000s the concept was taught and advocated by the communities. The demand for quantitative surveys to provide evidence for scientific observations, or hard data, went hand in hand with this.

Resilience strategies

In the years that followed, however, the indigenous population became somewhat uncomfortable about adapting these scientific methods, albeit by adding traditional knowledge. A certain frustration may have played a role here. After all, these frequently invoked hard facts have not had much effect in terms of global climate awareness. And a clear return to their own understanding and sociopolitical responsibility can be observed among the Canadian Inuit. The indigenous council ITK published the *National Inuit Climate Change Strategy* in 2019 following the vision that Inuit "… are prospering through unity and self-determination." It links climate risks to the social and economic inequality that has existed between the Inuit and the rest of the Canadian population for generations. The climate assessments of the early 2000s are now updated, the need for international action between the Canadian Inuit and Greenland Inuit has been recognized, and global models for indigenously managed protected areas have been developed. The considerations are differentiated and move along the most pressing problems caused by the changing sea ice and the associated fish and wildlife occurrences: the economic benefits that can arise from increased industrial activities, shipping traffic, and tourism are offset by the loss of huge areas and natural hunting resources. Previously unknown species that migrate to the warmer northern regions could be made accessible with modified hunting techniques; for example, by using boats. However, in any case, risks have to be taken into account, such as the increased dangers from melting pack ice or reduced food security, but above all the impacts on traditions, culture, and spirituality will lead to losses in cultural self-determination. The general danger for local societies is that traditional knowledge is no longer or no longer always appropriate (ITK 2019, 10, 14).

The feasibility and effectiveness of adaptation, moreover, hinge on having sufficient knowledge of the environment to travel and hunt in unfamiliar locations and on the flexibility of the wildlife management regimes that define the timing of seasonal hunts, as highlighted by Pearce et al. (2015). Regarding the aforementioned examples,

> adaptation to changing travel conditions on the land or ice or availability of wildlife may involve using alternative modes of transportation (e.g., a boat instead of a snow machine), hunting in unfamiliar locations, harvesting different species, or hunting at different times of the year. Here adaptation depends in part on the availability of capital resources (to purchase hunting equipment and supplies) or social capital (to borrow supplies) and time to participate in the hunt (e.g., constraints of wage employment).
>
> *(Pearce et al. 2015, 236)*

Therefore, adaptive capacity can be described as a set of resources that represent an asset base from which adaptations can be made.

The list of Inuit-driven climate initiatives covers ongoing environmental studies and resilience strategies for health, subsistence, knowledge, infrastructure, and energy.[8] There is intensive youth work to teach about huntable game and environmental variables, resource use, and food security measures. But the defined adaptation strategies also include programs such as climate efficiency and conversion to renewable energy. The local population is working on concrete programs and measures to support the necessary adaptations, which are also suitable for making an indigenous contributions to stabilizing the climate. Political demands on land and sea rights are also currently being expressed with great urgency in connection with the navigability of the Northwest Passage (Kuptana et al. 2018). Rights on land and at sea are of particular importance to indigenous peoples in view of national energy projects that now seem possible due to the melting of the polar ice caps. In the Arctic region, the Northeast and Northwest Passages in particular play a decisive role in the power struggle among the riparian states. Internationally, Canada considers the possible Northwest Passage Zone to be Canadian but the United States and Europe consider it to be international waters. Russia and China can also be assumed to be pursuing their own agendas here. The first seeks to expand the Northeast Passage or "northern sea route," and the latter envisions a polar silk road. The indigenous Arctic inhabitants are directly affected by this—hardly considered by the respective national governments. But the possible passage zones of the Northwest Passage are claimed by the Inuit as the "Inuit Sea." Corresponding map material documenting the traditional territorial extent and hunting areas of the Inuit was presented in the 1970s (Kuptana et al. 2018).[9] The IFA (Inuvialuit Final Agreement) of July 25, 1984, accordingly established land claims extending from the Beaufort Sea through the Amundsen Gulf to the Mackenzie Delta.

The Canadian Inuit are thus organizing themselves to push ahead with the necessary adaptations to global warming according to their own needs, experience, and knowledge. In the indigenous perspective, however, "traditional" knowledge should not be reflected solely in individual scientific perspectives. It may be comparable with the term "science," but the "Inuit way of knowing" sees humans in close connection with their environment and the universe. According to Kuptana and Napoyok-Short, "[...] outsiders have termed this knowledge base as 'traditional knowledge,' limiting the Inuit Way of Knowing to the past, a source viewed as anecdotal evidence and of little consequence for inclusion in discussions that impact Inuit in the Arctic. The term 'traditional knowledge' is not in keeping with the Inuit definition of the world around us" (2016). The spiritual practices of the Inuit are related to other dimensions. Stephen Hawkins, Kuptana continues, addresses the possibilities of other dimensions and time travel. This is very close to the view of the Inuit. In this way, when the first rocket landed on the moon, Inuit shamans had already been going there and back. What we can see here is the expression of "cultural capacity" with which the conditions for necessary adjustments are finally created rather than the pure economic or social "adaptive capacity."

Underlying concepts

But is a shaman's flight to the moon "factual"? In the physical sense, certainly no; in the spiritual sense, certainly yes. The ethnologist is extremely familiar with such questions. The mythic view of the world still exists among many indigenous peoples, but it is considered by rationalists as not real and therefore not existing, not provable, and, as a result, not factual, pure conjecture, or superstition, a compound of stories or a kind of literary description. Hence, in cross-cultural dialogue, it is very difficult to communicate eye-to-eye about knowledge if the knowledge comes from traditional empirical values and makes use of specific cultural ciphers and ideas.

We do not even need to attempt to go deep into the definition of spirituality and religiosity (which, by the way, could facilitate demarcation and acceptance from the point of view of rationality). We are concerned here with myths and nature; that is, with the way in which the "natural" reality of life is described as mythical space. Nature is animate environment. Humans, animals, the atmosphere, the Earth, and, finally, also the universe form the described space in the mythical world pictures, in which the known and the unknown, the beneficial and the dangerous, the living and the perished have their place and exchange. The crunching of the ice, the howling of the wind, striking landscape structures, the animals, the sun, as well as the moon and constellations of the night sky are assigned to mythological figures and their stories. An intentional or unintentional exchange with them is conceivable at any time. They determine the possibilities of daily life, but they also stand for the fear it can cause. The human being is only one building block. The mythical figures have their own will. To influence their powers positively for man is the goal of any spiritual effort.

As an illustration of this difficult discourse between rational and mythical worldviews, I will refer to MacDonald's account (1998) of an incident dating back to the 1980s. In Igoolik, government biologists wanted to limit the hunting of polar bears in the region and met with community representatives.

> One old hunter strenuously objected to their proposals, pointing out that polar bears had intelligence matching or exceeding that of humans and, as such, could be 'taken' only when they wanted to "give themselves". By way of explanation the hunter told of a time he followed fresh bear tracks. [...] The tracks suddenly ended, and there, on the tundra, was a rectangular block of ice. Clearly the polar bear, not wanting to be taken, had transformed itself into ice. The government biologists were bemused at this explanation, whereupon the old hunter told them that if they did not, or could not, believe him then they knew nothing about polar bears.
> *(MacDonald 1998, 18)*

But what about the Inuit definition of "the world around us" and "the spiritual practices related to other dimensions," which are rooted in daily life? In fact, the sun, moon, and constellations belong to the complex of oral traditions and thus to the store of life wisdom and applied knowledge. Beyond the importance of celestial and atmospheric spheres for the Inuit as a hunting people, "the Arctic sky held many of the metaphors and symbols shaped by the Inuit intellect to depict, embellish, and explain their distinctive world and its myriad realities" (MacDonald 1998, 1). Although a systematic attempt to portray astronomy is missing over the entire ethnographic period,[10] there is sufficient evidence of well-founded knowledge and its abstraction in cosmological concepts and its concrete application to support orientation and navigation. However, traditional knowledge is characteristically personal; its acquisition and application in varying degrees depend on circumstances in communities, families, and individual capacity. Naturally, the ethnographer's or rapporteur's attention to certain topics also depends on his personal preferences. Robert Peary, who himself was a surveyor and navigator, was quite impressed by the astronomical knowledge of Northwest Greenlanders at the end of the 19th century and even assigned some of the stellar examples to Western nomenclature. Almost 100 years earlier, the theologian and missionary David Cranz spoke of speculative and romantic stories with which the Inuit served the Europeans in order to please them or even to test their judgment and credulity (MacDonald 1998, 2–6).[11] The Sun and Moon were often reported by ethnographers in the 19th and 20th centuries as a pair of siblings, with the Moon appearing more significantly in the more northern polar regions than in the more southern inland areas. This is not surprising given the polar night. Among Netsilik Inuit such as the Caribou Inuit in the 1920s, Rasmussen (1931, 17) perceived the moon as rather vague, although it appeared to have some influence on women's fertility and hunting success. Spiritual strength and special power can be seen in stories in which weak and neglected protagonists seek far too great a challenge but finally overcome all dangers, as recorded by various folktales, by choosing and killing an extremely large walrus instead of a smaller and less dangerous one, for example (Rasmussen 1931, 231, 233).[12] With the spiritual journey, the shaman or *angakoq* proves his ability to overcome the dimensions of space and to be at eye level with the powers of the universe. He talks to them, visits their houses, and hunts and eats with them.[13] The close interweaving of existence and the universe is still part of the Inuit's self-understanding and one of the foundations of their view of the world, as Kuptana and Napoyok-Short emphatically emphasize: the knowledge of the Inuit "[…] has its roots and principles in the belief that the world around us and life itself are interdependent, interrelated, inter-dimensional, multi-disciplined, interconnected, intergenerational, evolving, and holistic" (2016). Thus, the isolated use of expedient examples without their reference to the contemporary and traditional body of knowledge is criticized as an inadmissible restriction of the "Inuit way of knowing." and Ian Mauro and Zacharias Kunuk (2010) took a critical look at "external" interpretations a few years before this statement with the documentary *Inuit Knowledge and Climate Change.*

Ian Mauro, as an environmental scientist, geographer, and filmmaker, specializes in climate change communication and community-based research. He has been engaged for years with Inuit communities in social and ecological science projects.[14] Zacharias Kunuk is an award-winning filmmaker who, in 1990, founded Igloolik Isuma Productions, the first Inuit film production company, together with Norman Cohn as cameraman, Paul Apak Angilirq as editor, and Pauloosie Qulitalik as cultural narrator. Isuma TV is a project in support of more self-determination and against the loss of the Inuit's own culture. Kunuk asks: "Can Inuit bring storytelling into the new millennium? Can we listen to our elders before they all pass away? Can we save our youth from killing themselves at ten times the national rate? Can producing community TV in Igloolik make our community, region and country stronger? Is there room in Canadian filmmaking for our way of seeing ourselves?" It is "[…] to show how our ancestors survived by the strength of their community and their wits, and how new ways of storytelling today can help our community survive another thousand years" (Kunuk 2017).

The aim of the filmmakers is a "participating" visual investigation of the Inuit's observations on current climate change—completely from their own perspective and with recourse to personal life memories. Although its content follows on from the previously mentioned contributions from 2000–2001, it allows the interviewees to speak for themselves and in their own language. Mauro said: "[…] we made a film and people are telling their own perspectives, and their own stories, and that really inverses the dominant way of doing research. […] Instead of this ivory tower approach, this is a democratic way that makes the research make sense to people, because it's just *storytelling*."[15]

In German, the term "storytelling" does not exist or, rather, exists in another context.[16] "Stories" are classified according to content criteria and their contexts as "myths, legends, or fairy tales." In German-language texts, it is therefore common to use "storytelling" without translation. The term "narrative" is used in the same way. In English, it belongs to the larger complex of "oral traditions" and therefore refers more to speech performance. "Storytelling" always took place in a communication situation, and in previous oral cultures the presence of the audience ensured the quality of content and reproduction. In this way, the form of representation should be emphasized, which is so tremendously important for the specific narrative culture. The Inuit narrative "[...] depends heavily on vocal and visual mannerisms to characterize the actors taking part in the story and to increase the audience's appreciation of the action" (Hall [1939] 1975, 412).[17] Oral traditions in oral cultures are closely interrelated with other forms of expression, such as body language, intonation, sound formation, and pauses in speech to emphasize dramatic effects. These overall expressions are not adopted by translation into writing but are largely excluded. The "oral texts" are the equivalence to the written but not always the essence of oral culture.[18]Ingold (2022, 196–197) raises the question: "Suppose that we read a text silently. [...] We are just reading the words; nothing more. Can we nevertheless inhabit the book's pages as we inhabit the land?" Thus, he feels that literature, as art, would be needed to draw the forces of physical environment.[19]

In contemporary discussions, it has become customary to use the term "storytelling" with great freedom and without a precise definition of its application in different speech situations or in its already-written form. Oral traditions thereby grow together in writing, become literary artefacts, and often remain strangely unreflective in discussion. Thus, they are subject to an inappropriate narrowing down of their application in formerly oral cultures—both yesterday (history) as well as today. The spectrum of oral cultures—as far as we can still trace it at all—is rich in the most varied forms of speech.

There were the moral speeches of the elders, speech rituals for special life situations, for hunting or festive occasions, religious speeches, declarations, betting speeches, or the sung word in song dances, etc. Ethnographic compendia generally attempt a thematic categorization of narratives that reflect well their range of content but necessarily neglect narrative function and situation.[20] Today, with written culture having long since taken hold among the Inuit and being only one example among many of the profound changes in their cultural experience, the collection of narrated memories is an important building block of Inuit cultural memory. "It is a tradition among Inuit to pass down wisdom and knowledge and the best way to do that is to share stories," introduces Mary Powder (Desjardins 2020).[21] "We did not have pen and paper. We needed knowledge for our survival and to pass the time."

The term storytelling as an everyday means of understanding must necessarily be delimited from mythological concepts that religiously and spiritually underpin the life-world and even from other examples of more ritualized speeches, such as the well-known battle of words.[22]Spalding (1979, v) attests to the highest quality of the myths recorded in the 1950s in northern Alaska, and although "[…] the banal and the profane is raised to the level of the special and miraculous […] they come bathed from a depth of the human spirit where […] the most important and mysterious spiritual quests and battles have their rise and fulfillment." In most cases, "storytelling" hides the larger underlying stock of generic knowledge and cultural convention. If we were to apply Habermas's distinction between "performative consciousness" and "fallible knowledge," there exists a mode of "background knowledge that runs with us, that is intuitively certain but remains implicit, that accompanies us in our daily routines. […]" Certainties that are present, as it were, like amalgam, and which, when disturbed, "turn into an issue at any time […]." When dissonances occur in life-worldly certainties,

> they can only be put into propositional form with the loss of their performative mode. […] For what we know in this intuitive way, we can only make explicit by transforming it into a description; in the process, the consummation mode of the merely known dissolves—it disintegrates, as it were.
>
> *(Habermas 2019, 466–467, author's translation)*

Ingold underlines a "mental domain" that resists explication and manifests itself in habitual practice—though not in distilled form: "The greater part of what we know, we cannot explain." Thus, "[…] storytellers are wayfarers […] they need to attend to things as they go, to recognize subtle cues in the environment and to respond to them with judgment and precision" (2022, 227–233).

The recognition and description of the changing living environment, referred to as "storytelling" in the film documentation, represent traditional and remembered framework conditions supplemented by experienced personal knowledge. In the interview situation, the appropriate information will be a matter of factual circumstances, social memories, personal preferences, or generally accepted statements on the subject. There is another phenomenon in cross-cultural conversation that I would call "communicative appropriation." This consists of the use of knowledge, representations, or motifs that are familiar to the interlocutor because they come from his or her own cognitive environment. In this way, understanding can be achieved but, above all, a conversation at eye level is possible.

I would like to provide two examples that could be described as communicative appropriation, because although they arose out of a certain situation and provided an indication of an underlying complex of knowledge, they did not reflect this complex of knowledge at that particular moment: among the northern Alaskan Inuit, the story goes that the first missionaries who came to teach God's biblical creation of the world were countered with: "Very well, God made the world, but Raven made it first" (Spencer, cited in MacDonald 1998, 23).

However, creation reports are rather fragmentary in Inuit mythology recorded during the ethnographic period. They refer more to the interpretation of natural phenomena than to the creation of the world. The concept of creator and cultural hero alluded to in this example goes back to the influence of the neighboring Athapascan-speaking population and has Asian-Siberian reminiscences[23] from a cultural–historical point of view. In this way, however, a creation idea could be held against the missionaries. The next example takes us almost 200 years back into the past: When Hans Egede, the Danish-Norwegian missionary in Greenland at the end of the 18th century, was teaching the biblical catastrophe of the Flood, a listener told him that "[…] the sky before this rested on the tops of the mountains but after the great flood that flooded the whole earth, it was lifted up to the height it has now […]" (Egede 1790, 167). However, there is no evidence that the Greenlandic creation mythology knew about a flood. But the worldview of the Inuit from Alaska to Greenland took the Earth as the fixed center of the universe. It is surrounded by different celestial spaces that lie on top of each other like layers, each of which represents a world of its own. The vault of heaven with its celestial bodies moves above the Earth. The pillars that hold these arches are not arranged centrally but rather similar to the arrangement in a tent. At each corner of this flat, uninterrupted expanse of land and water—the Earth—there is a wooden column that holds another unlimited expanse—the sky (Jenness [1922] 1970, 190; Rasmussen [1932] 1949, 193; Hultkrantz 1981, 15; MacDonald 1998, 23–24). Yet, in the described discussion situation, a conception of a world renewal after the Flood was developed. This can only be explained and classified in context, and the knowledgeable reader will interpret this accordingly.[24] The speaker seeks some common ground for an exchange and intelligent discussion with the interlocutor.

Filmic statements

Let us consider the 2010 documentary by Mauro and Kunuk for a moment and its most striking element, the shift of the Earth's axis. Ian Mauro underlined it in an early interview: "[…] air pollution and CO_2 emission might not be the full story. […] It's quite fascinating to see the links between—you know—elders on their lands. I've literally been emailing with NASA scientists about the observations that Inuit have around the changes to the sun, the stars and the position of the earth on its axes."[25] To the Inuit elders who were interviewed, the assumption that the universe has shifted is recognizable by the changed positions of the sun, the stars, and the Earth on its axis.

Among numerous quotations, I would like to mention two extracts that circulate on the internet.[26] They highlight mainly the shift of the Earth's axis whereby the appropriate film sequences in both examples remain unchanged. The first film excerpt was posted by the renowned London institute Forensic Architecture, naming the original source as "[…] new documentary, the world's first Inuktitut language film on the topic, [which] takes the viewer 'on the land' with elders and hunters to explore the social and ecological impacts of a warming Arctic. This unforgettable film helps us to appreciate Inuit culture and expertise regarding environmental change and indigenous ways of adapting to it." Although referring to Mauro's explanation of the close connections of the people with their country and the socioecological effects of climate change, above all, the new experience of disorientation in these endless landscapes of snow, ice, and water, in the selection of visual material also focuses on the assumption that the Earth's axis is shifting. In the second example, the corresponding sequences are anonymously posted on the Internet. The author has arbitrarily added a dedicated title page (*Pole Shift? Observation of the People*) to the film clip and has chosen a snowmobile traveling scene for the first sequence.

To go more in depth into some aspects of spoken words, remembered living conditions, and visualized events, brief extracts of the text are given below. The interlocutors remember travels with the dog team and invoke atmospheric knowledge and methods of orientation and navigation that now need to be adjusted.[27]

We'd go to the floe edge by dog-team
Leaving early in the morning
We had to arrive at daylight in order to catch seals
We had one hour of daylight
Today we have a two hour window to shoot
I've lived here all my life and have also watched the sun
Where it rises has not changed much, but the sunset has shifted way over
Perhaps the earth has tilted on its axis
The earth has changed its tilt.
I don't know exactly when it happened
But I know the sun used to set close to the highest mountain peak
After the shift, the sun now sets past the highest peak
There is hardly any tongue drift these days
Tongue drifts are directional markers used for travel.
They are formed by North Wind, which has changed.
I learned to observe the ground, if stars are not visible
Today, stars also look different
At night, returning from a hunt using the stars
It's noticeable they are no longer in their proper positions
Our world has changed—land, sky and environment
Tongue drift now point a different direction
When moving east, we crossed them sideways
Today heading east, we go with the drifts
The shifting wind has caused this
...

Obviously, the filmic scene shows by no means travel with a dog team but travel with a snowmobile, and its main focus, the "tilted Earth on its axis," seems to be some transferred natural scientific element. Traditionally, there is no evidence that there was an Earth axis in the sense of an *axis mundi* in the worldview of the Inuit. The sky arch is, as already mentioned, supported at the respective corners by pillars. On the contrary, there is at least one indication that the existence of an Earth axis was doubted only recently. In the 1990s, when Spalding was in Repulse Bay talking to an interlocutor about views of the universe and discussing the Sun, the Moon, the Earth, and the celestial bodies, he came down to Earth, talking about its rotation around the Sun and about its own axis. His interlocutor found this unlikely and replied, "I think it's very unlikely that this thing moves though!"[28] The theory related to the Earth's axis is thus obviously a new element.

But the really important thing about these statements is the competence to think in large, universal contexts. Heavenly and atmospheric spaces were of paramount importance for the Inuit as they were for all hunter peoples. Seasonal and daily time structures were detected by means of known celestial bodies, and atmospheric conditions were reflected in weather patterns. Statements like "Our world has changed, land, sky and environment," "today the stars also look different," "wanted to talk about these changes to the sun and the environment," "We get heat from the sun after our world tilted," "if our world has shifted or it has tilted north, that's what I wonder about" refer to the concepts of *Sila*, the Earth with its atmosphere, the air, and everything that appears in it, as well as *Qilak*, the heavenly sphere,[29] which are still anchored in the reality of life. Knowledge of astronomy and the observation of astronomical processes and movements are also essential for orientation when the Sun cannot be seen. However, these "wayfinding practices" were also oriented to land markings or snowdrifts, the stability of which, in turn, depends on wind conditions. The Norwegian Christian Leden (1927, 134–138)[30] described this impressively when he advanced "over Keewatin's ice fields" into the area north of Hudson Bay during his expedition in the 1920s. *Keewatin* is an Ojibwa word and means "north wind." In the course of a snowstorm, he and his companion lost their orientation. They decided to let the dogs relax and follow them in the hope that they would run to the next settlement and thus to the next feeding place. However, because of clumsiness, the dogs tore loose and escaped. In dense snowfall and impenetrable darkness, the only way for both travelers to find their way around was the northeast wind, which "normally lasts at least one day." They walked so that they had the wind "half from the front and half from the left" and finally found the next inhabited camp site. Navigating in this vast expanse of water, ice, and differently shaped snow landscapes, orientation in daylight, at night, and especially during the polar night was essential for survival. Birket-Smith (1929, 154) recorded "[…] an outstanding ability to orientate themselves by means of a number of quick and scarcely conscious decisions at to the position of celestial bodies, the meteorological state and so on."

This example gives an idea of the drastic significance of the film's report "[...] weather was predicted by observing the sky. Cloud formations indicated wind direction. Now it is different. First, they form a way; then they quickly change, telling you a different story" (Mauro and Kunuk 2010, 00:59). Snow drifts as reliable indicators of direction are therefore a thing of the past. But it is not only about simply equipping all hunters with GPS technology. The way of life requires that weather be predictable, which has not happened for at least a generation. Experience and competence cannot be applied anymore. Hunting strategies—or, more generally—nutritional strategies must be adapted, and complete movement profiles are changing.

Historical impulses

Let us remain with overland travel for a moment. In the 1920s, the dog team was still the most important and reliable means of transport. Researchers also used it. As is well known, Rasmussen became a master among dog sled drivers. But dog teams are expensive; they eat a lot of meat. Originally (i.e., during the *pre-contact period*), a dog team consisted of only two or three dogs.[31] The whaling industry then brought lasting change. Even before the Hudson Bay Company went north, there was trade with the northern regions. Europeans were dependent on whale oil. In the late 1700s and 1800s, Scottish and American whalers had already reached Pond Inlet and Cumberland Sound on Baffin Island.[32] In the early 1860s, they found their way to the rich whale grounds of the even more northern regions. Because the Inuit were whalers long before that, they joined the Western whalers, who welcomed them as hunters and navigators. The Inuit supplied fresh meat, ivory, and fur in exchange for knives, needles, and clothing. But from the Western whalers they also brought home new forms of song and dance, pregnant women, and previously unknown diseases. Clans mixed together like never before, and traditional migration routes were interrupted. The introduction of rifles increased the dependence on traders for ammunition and led to a higher yet predictable hunting rate, which provided sufficient food. The surplus of food could now be fed to the dogs. Larger dog teams were able to travel longer distances and covered a larger area. Finally, dogs were considered a status symbol. However, the need for larger quantities of meat again led to greater pressure on hunting success and to even greater dependence on the traders. The whaling industry thus had a great influence on people. This experience was also decisive for further adaptation strategies, although it is usually neglected in the literature (McGrath 1984, 5–6; Matthiasson 1992, 32–33). When industrial whaling experienced its first major slump in the early 20th century, the Inuit were no longer able to return to their simpler way of life. However, the gap was quickly filled by fur traders, who opened trading posts in the coastal regions.[33] In turn, the emergence of commercial settlements resulted in the local centralization that exists today. From the 1920s onwards, all Inuit had relatively easy access to trading posts, and in the 1960s there was mass migration from the camps to the settlements. This also brought the people under the directives of the political leadership as well as federal and territorial bureaucracy (McGrath 1984, 6; Matthiasson 1992, 24–28).

The phase of relative prosperity because of trade and hunting safety during the whaling period; the recognition of indigenous competencies, especially in the field of hunting and navigation; and the simultaneous relative independence and freedom of movement still live on strongly in cultural memory. In particular, the frequent long and slow overland journeys with the dog teams and the whole family, the camping stops during the journey, and waiting until the igloo was finished helped to pass on an understanding of astronomy from parents to children.[34] Yet today, traveling by snowmobile means the opposite: there is little time left to contemplate the vault of the sky. The hunters always drive similar routes, and other members of the family accompany them mostly in spring when the sunlight obscures the stars. Some elders also noted that the light from the settlement's streetlamps leads to "light pollution." Elders on MacDonald's research team reported that their understanding of astronomy was rather low compared to that of their parents and grandparents. There was thus once again a relatively rapid thinning out of this knowledge (MacDonald 1998, 6–7). Thus, larger movement radii through larger dog teams created the prerequisite for deepening knowledge of astronomy as a navigational instrument, a skill that is no longer needed to this extent today. Consequently, the Canadian strategy for national Inuit climate change focuses on the main issues where in many cases "traditional knowledge" is no longer sufficient and its impact on traditions, culture, and spirituality will lead to losses in cultural self-determination (ITK 2019).

In comparatively few years, we have become contemporary witnesses of great changes. But at the same time, we can also trace the far-reaching changes that are still revealing themselves from the past. The examples are meant to show that interaction in cultural processes is the standard and not the exception.

Conclusion

But still one question remains: has the Earth's axis actually shifted? In general, it is a matter of masses. Currently, there are different hypotheses.[35] One reason is the continuing continental drift. The displacement of the Earth's masses, which is taking place in very little incremental steps, can also be proven and measured. Around the year 2000, however, there was a larger "movement." According to one theory, this was a result of the melting masses in the polar region as well as in Greenland and in Antarctica. However, this melt must be seen in connection with the simultaneous loss of water mass in various terrestrial areas that has been caused by dry periods as well as the massive extraction of ground and surface water. Melting ice due to incompatible CO_2 emissions and disproportionate water withdrawal led together to a loss of mass that is likely to cause movement of the Earth's axis. In a most recent study analyzing comparative data from 1981–2020 and 2002–2020, researchers from the Chinese Academy of Sciences led by Shanshan Deng (Deng et al. 2021) show that the accelerated decline in terrestrial water reservoirs—due to melting and groundwater pumping—is thus the main cause of the rapid eastward polar shift since the 1990s.

Indeed, we seem to be on the brink of a great change, if not on the edge of an abyss. Therefore, we should do what is factual, based on evidence, and reasonable and simply what we can do right now: namely, to use water carefully, reduce man-made CO_2 emissions, and root this in our lifestyle and existence. The Inuit's competence in using a possible—in Western understanding—geophysical condition as an explanation for global warming suggests a close familiarity with stellar constellations and astronomical conditions. In traditional Arctic mythology, star lore and the figurative design of the night sky are deeply rooted. It is knowledge that had to be compulsorily developed for orientation and navigation during hunting expeditions and overland journeys, especially during the polar night. It is slowly disappearing among the younger generations, who, equipped with technical means, also have changed hunting behavior. It is therefore comprehensible that the thesis of the Earth's axis shift was voiced by the Inuit elders on the basis of a traditional knowledge of celestial bodies and supported by clear observations of environmental conditions. The adaptation of the Earth's axis as a "modern" element to explain the massive changes that have impacted the lives of the polar population within a few decades—not to say within just one generation—underlines that a figuratively very strong motif was needed to make the dramatic nature of the changes clear.

In essence, the demand for "climate justice" must be a demand for "existential justice." Losses of cultural self-determination cannot be quantified. Taking responsibility for global warming is not only a financial task but also a humanistic challenge for every industrial nation and those that merely exploit or consume fossil resources.

Notes

1 Translated by the author; regarding changes in subsistence—for example, reindeer hunting—see also Hall ([1939] 1975, 26–29).
2 Internet hits for the keywords "Earth axis" and "Inuit" are so numerous that research concerning their origination is time-consuming. Moreover, the motif is widespread in the Mackenzie Delta area, for example (Dr. Franz Krause, University of Cologne, personal communication).
3 The scientific committees of the World Meteorological Organization became internationally networked with the founding of the IPCC (Intergovernmental Panel on Climate Change) in 1988 and publish annually on the basis of comprehensive and international data collections. See, inter alia, the landmark report of 2007: Mitigation of Climate Change (IPCC 2007). At the political level, this was

followed by the Kyoto Protocol of 2005 and the Paris Climate Convention in 2015 (cf. Archer and Rahmstorf 2010, 1–15).

4 Inuit broadcaster, author, rights activist, and former president of ITK, in: Canadian Encyclopedia, keyword "Rosemarie Kuptana."

5 ITK: Inuit Tapiriit Kanatami. Representative of the majority living in Inuit Nunangat, the Inuit homeland encompassing 51 communities across the Inuvialuit Settlement Region (Northwest Territories); Nunavut, Nunavik (Northern Quebec); and Nunatsiavut (Northern Labrador). Inuit Nunangat makes up nearly one third of Canada's landmass and 50 percent of its coastline.

6 At min/sec 38:23, in *Sila Alangotok: Inuit Observations on Climate Change* (Dickie and Woolf 2000); see also Sutherland (2003).

7 For further definitions of "TK" and "TEK," see Pearce et al. (2015, 235, 240).

8 The ITK (2019, 5–18) lists the respective examples of Inuit-driven climate initiatives.

9 This map was created by Inuit hunter Henry Hokshun of the Goja Haven Region in the 1970s. The data were drawn from seven decades of travel in the region as part of the Inuit land use and occupancy project to help show where their traditional lands are. It is stored in the Library and Archives Canada.

10 Referring to the main contributions of the 19th and 20th centuries. Historically there might be different perspectives to structure the various phases of the Inuit–Western approach. For instance, Matthiasson (1992, 24–25) differentiates between the "pre-contact period," "whaling period," "contact-traditional period," "centralization period," and "becoming Canadians."

11 The author's compendium finally was based on various interviews with several "well-informed Inuit elders" from Igloolik (NWT) in the years 1988–1997.

12 Mythical tales that deal with extraordinary deeds or creations often give neglected, abused, or orphaned boys the role of "epic heroes." The deeper understanding is the particular strength demanded of orphans who, without parents, were often at the mercy of the willingness but also the discretion of neighbors or relatives in a very harsh reality of life.

13 In his compilation, MacDonald (1998, 263–288) mentions various authors of different ethnographic periods like Boas, Hawkes, Holm, Holtved, Nelson, and Rink. The spiritual journal plays a role in early ethnographic reports from the 19th and 20th centuries, like Boas, as well as in later ones, like MacDonald. According to examples given to us by Rasmussen (1922, 22–34), necromancers can perform ghost flights to the land of the dead in heaven, to the land of the dead in the underworld, or to the moon.

14 https://www.isuma.tv/members/ian-mauro. Accessed August 15, 2022.

15 CBS News February 4, 2011. Mauro's use of filmmaking in academia has been controversial with some former faculty members: "With past films that I've made, there has been resistance from the university community specifically about whether or not filmmaking is actually research," said Mauro.

16 The "storyteller" or "telling stories" is often connoted in a positive sense with imaginative speeches but also in a negative sense simply with the untruth.

17 Hall, moreover, refers to Gidding's report on the Kobuk Inuit and adds that "gestures, pauses, and voice modulation are used continuously for emphasis and dramatic effects." The importance of an acting representation is underlined with Spencer's (1959, 383) description of Point Barrow storytellers, who reports "[…] good mimicry was much applauded as the tale itself."

18 Münzel and Scharlau (1986, 163–164), generalizing the issue to translate and write down indigenous oral traditions.

19 Ingold refers to Fredrickson's (2015) unpublished doctoral thesis "Orientation in Weather: A Northern Textual Ecology," Department of English and Film Studies, University of Alberta.

20 For example, "Myth and Legends of the Origin," "Fables and Tale of the Animals," "Historical Legends," "Human Beings," "Epic Tales," "Manslaughter and

Revenge," "Magic," etc. (cf. Holtved [1951], as only one example among many others). Rasmussen (1922, 6) attempts to incorporate the Inuit view: there are epic, religious, humorous, and soporific stories. We can imagine the narration situation when the soporific ones were described to pass the time and winter and put the listener to sleep (the greatest praise was felt when the listener did not finish the story).

21 A new podcast featuring "traditional Indigenous storytelling" of the CBC archive in the Inuktituk language: https://www.cbc.ca/news/canada/north/cbc-indigenous-archiving-project-1.5155932. Accessed August 3, 2022.

22 Rasmussen ([1922] 1991) recorded the so-called Minstrel Contest or "Sängerkrieg" in Greenland, a competition among the best Greenlandic poets and singers. The duels were so called by Rasmussen in borrowing from the European Late Medieval minstrel contests. The competition is a power struggle for intellectual presence and "world knowledge." The struggle for spiritual prestige and power goes hand in hand with this.

23 See epilogue by Anne Schmücker in Rasmussen ([1932] 1949, 194); Hatt (1949, 5); Nelson ([1899] 1983, 425–426); Hultkrantz (1981, 15). The infinite and perilous expansion of the living world is acknowledged in various oral traditions; here, Rasmussen ([1932] 1949, 148–156) and Nelson ([1899] 1983, 499–505) are only mentioned as examples.

24 A systematic attempt to understand the migration of European biblical narrative elements to North America, provided by Thompson (1919, 452), remained rudimentary and unfortunately is still not worked out. Most probably it is too late now. Concerning the most obvious motif, Noah's flood, he recorded evidence at Thompson River, Lillooet, and Tepecano. Rooth's investigation *The Raven and the Carcass* (1962) ventured extensive pursuit of the selected motif in Europe and Asia and supplemented North American Deluge Myth examples.

25 CBC News, December 10, 2009.

26 *Inuit Knowledge and Climate Change*, extracted by Forensic Architecture (2012), and *Pole Shift? Observation of the People*, extracted by "Anonymous" (n.d.), both based on Mauro and Kunuk (2010).

27 Mauro and Kunuk (2010) at min/sec 42:01 and following.

28 However, one cannot derive "world picture dogmatism" from this (Spalding, cited in MacDonald 1998, 24). MacDonald (1998, 6), who has compiled a compendium of Inuit astronomy and the corresponding myths on the basis of ethnographic sources in collaboration with 27 Inuit experts, points out that traditional knowledge is characteristically personal and that its acquisition and application in the various forms and depths depend on communities, families, and individuals. The elders often made it clear that they speak for themselves, that their opinion is not necessarily correct in an absolute sense, and that other elders may have a different opinion.

29 Thus also quite concretely the weather, winds, the aurora, and other natural phenomena (cf. MacDonald 1998, 1, 35).

30 The Norwegian Christian Refsaas studied in Berlin and adopted a German name during this time. He led several expeditions, which were accompanied by collecting activities for museums as well as music–ethnological research. In the 1930s, he befriended fascist and racist organizations in Norway and Germany and subsequently supported national socialist propaganda.

31 In the 1920s, Leden was already underway with six dogs.

32 In 1590, according to Keith Crowe (1974, 105, 350) whaling ships were underway near the pack ice between Greenland and Baffin Island.

33 In the chapter "The Almighty Furrier," Leden (1927, 46–47) impressively describes how much power the trading stations exercised.

34 Considering these kinds of correlations, it seems to be doubtful that "[…] Inuit now travel and hunt often by all-terrain vehicle (ATV), snow machine, and boat than by dog team and use high-powered rifles instead of bow-and-arrow or spears,

but the objective of their efforts remains unchanged and traditional ways of knowing continue to be relevant" (Pearce et al. 2015, 235). The impacts are different and affect social and economic life.
35 A representative article with respect to the Earth axis topic and the mentioned theories can be found on the NASA Jet Propulsion Laboratory homepage (2016).

References

Archer, David, and Stefan Rahmstorf. 2010. *The Climate Crisis. An Introductory Guide to Climate Change.* Cambridge: Cambridge University Press.

Ashford, Graham, and Jennifer Castleden. 2001. *Inuit Observations on Climate Change—Final Report.* Winnipeg: International Institute for Sustainable Development (IISD).

Birket-Smith, Kai. 1929. *The Caribou Eskimos—Material and Social Life and Their Cultural Position: Vol. 5. Report of the Fifth Thule Expedition 1921–24.* Copenhagen: Gyldendalske Boghandel.

Crowe, Keith. 1974. *A History of the Original Peoples of Northern Canada.* Montreal: Arctic Institute of North America, McGill-Queen's University Press.

Deng, Shanshan, Suxia Liu, X. Mo, and Peter Bauer-Gottwein. 2021. "Polar Drift in the 1990s Explained by Terrestrial Water Storage Changes." *Geophysical Research Letters 48: 1–10.*

Desjardins, Lynn. 2020. "Traditional Inuit Storytelling Heard on New Podcast." Accessed August 3, 2022. https://www.rcinet.ca/en/2020/10/21/traditional-inuit-storytelling-heard-on-new-podcast.

Dodds, Klaus, and Nuttall, Mark. 2019. *The Arctic. What Everyone Needs to Know.* Oxford: Oxford University Press.

Egede, Hans. 1790. *Nachrichten von Grönland.* Kopenhagen: Bey Christian Gottlob Prost, privilegirten Universitätsbuchhändler.

Habermas, Jürgen. 2019. *Auch eine Geschichte der Philosophie: Band 1. Die okzidentale Konstellation von Glauben und Wissen.* Berlin: Suhrkamp.

Hall, Edwin S., Jr. (1939) 1975. *The Eskimo Storyteller. Folktales from Noatak, Alaska.* Knoxville: The University of Tennessee Press.

Hatt, Gudmund. 1949. *Asiatic Influences in American Folklore.* Kopenhagen: Munksgaard in Komm.

Holtved, Erik. 1951. *The Polar Eskimo. Language and Folklore.* Kopenhagen: C.A. Reitzels.

Hultkrantz, Åke. 1981. "North American Indian Religions in a Circumpolar Perspective." In *North American Indian Studies*, edited by Peter Hovens, 11–28. Göttingen: Edition Herodot.

Ingold, Tim. 2022. *Imagining for Real. Essays on Creation, Attention and Correspondence.* New York: Routledge.

Intergovernmental Panel on Climate Change (IPCC). 2007. *AR4 Climate Change 2007: Mitigation of Climate Change. Contribution of the Working Group III to the Fourth Assessment Report of the Intergovernmental Panel on Climate Change.* Cambridge University Press. Accessed February 26, 2023. https://www.ipcc.ch/report/AR4/.

Inuit Tapiriit Kanatami (ITK). 2019. *National Inuit Climate Change Strategy. Report of Inuit Tapiriit Kanatami.* Ottawa: Inuit Tapiriit Kanatami. Accessed August 11, 2022. https://www.itk.ca/.

Jenness, Diamond. (1922) 1970. *The Life of the Copper Eskimos. Part A of Volume XII. A Report of the Canadian Arctic Expedition 1913–1918.* New York and London: Johnson Reprint.

Koester, David. 2002. "When the Fat Raven Sings: Mimesis and Environmental Alterity in Kamchatkas' Environmentalist Age." In *People and the Land. Pathways to Reform in Post-Soviet Siberia. Siberian Studies*, edited by Erich Kasten, 45–62. Berlin: Reimer.

Kunuk, Zacharias. 2017. "The Inuit Art of Storytelling." Accessed August 11, 2022. http://www.isuma.tv/isuma/isuma-style.

Kuptana, Rosemarie, and Suzie Napoyok-Short. 2016. "Inuit Ilitquisa: Inuit Way of Knowing." In *Arctic in Context*. Seattle, WA: The Henry M. Jackson School of International Studies, University of Washington. Accessed August 11, 2022. https://jsis.washington.edu/aic/2016/06/15/inuit-ilitqusia-inuit-way-of-knowing/.

Kuptana, Rosemarie, Suzie Napoyok-Short, and Inuit Qaujisarvingat. 2018. "The Inuit Sea." Arctic Focus. Accessed August 11, 2022. https://www.arcticfocus.org/stories/inuit-sea/.

Leden, Christian. 1927. *Über Kiwatins Eisfelder: Drei Jahre unter kanadischen Eskimos.* Leipzig: Brockhaus.

MacDonald, John. 1998. *The Arctic Sky. Inuit Astronomy, Star Lore, and Legend.* Toronto: The Royal Ontario Museum and the Nunavut Research Institute.

Matthiasson, John S. 1992. *Living on the Land. Change among the Inuit of Baffin Island.* Peterborough: Broadview Press.

McGrath, Robin. 1984. *Canadian Inuit Literature: The Development of a Tradition. No. 94. Canadian Ethnology Service Paper A Diamond Jenness Memorial Volume.* National Museum of Man, Mercury Series. Ottawa: National Museums of Canada.

Münzel, Mark, and Birgit Scharlau. 1986. *Quellqay. Mündliche Kultur und Schrifttradition bei den Indianern Lateinamerikas.* Frankfurt am Main: Campus.

NASA Jet Propulsion Laboratory. 2016. "NASA Study Solves Two Mysteries about Wobbling Earth." Accessed August 3, 2022. https://www.jpl.nasa.gov/news/news.php?feature=6332.

Nelson, Edward W. (1899) 1983. *The Eskimo about Bering Strait. XVIII. Annual Report of the Bureau of American Ethnology.* Washington, DC: Smithsonian Institution Press.

Pearce, Tristan, James Ford, Ashlee Cunsolo Willox, and Barry Smit. 2015. "Inuit Traditional Ecological Knowledge (TEK), Subsistence Hunting and Adaptation to Climate Change in the Canadian Arctic." *Arctic* 68, no. 2: 233–245.

Rasmussen, Knud. 1922. *Grönland-Sagen.* Berlin: Gyldendal.

Rasmussen, Knud. (1922) 1991. *Der Sängerkrieg. Eskimosagen aus Grönland.* Berlin: Zerling.

Rasmussen, Knud. 1931. *The Netsilik Eskimo. Social Life and Spiritual Culture. Report of the Fifth Thule Expedition 1921–24.* Copenhagen: Gyldendal.

Rasmussen, Knud. (1932) 1949. *Die Gabe des Adlers. Eskimoische Märchen aus Alaska.* Translated by Änne Schmücker. Frankfurt: Schütte.

Rooth, Anna Brigitta. 1962. *The Raven and the Carcass. An Investigation of a Motif in the Deluge Myth in Europe, Asia and Nord America.* Helsinki: Suomalainen Tiedeakatemia.

Ross, Sonja. 2000. "Mythology as an Indicator of Cultural Change. Hunting and Agriculture as reflected in North American Traditions." *Anthropos* 95: 433–443.

Spalding, Alex. 1979. *Eight Inuit Myths Inuit Unipkaaqtuat Pingasuniarvinilit.* National Museum of Man, Mercury Series. Ottawa: National Museums of Canada.

Spencer, Robert F. 1959. *The North Alaskan Eskimo. A Study in Ecology and Society.* Washington, DC: Smithsonian Institution. Bureau of American Ethnology Bulletin 171.

Sutherland, Dawn. 2003. *A Teacher's Guide for the Video, Sila alangotok: Inuit Observations on Climate Change: A Resource for Senior 2 Science.* Manitoba: Manitoba Education and Youth, School Programs Division. Designed for use with the video titled: Sila alangotok: Inuit *Observations* on *Climate Change.*

Thompson, Stith. 1919. *European Tales among the North American Indians. A study in the Migration of Folktales.* Colorado Springs: Colorado Springs College Publications.

Documentaries

Dickie, Bonnie, and Terry Woolf. 2000. *Sila Alangotok: Inuit Observations on Climate Change.* International Institut for Sustainable Development (IISD), 42:15 min. https://www.youtube.com/watch?v=nOZ5LLA-SKg-.

Mauro, Ian, and Zacharias Kunuk. 2010. *Inuit Knowledge and Climate Change.* A production of Igloolik Isuma Production & Kunuk Cohn Productions, 54:07 min. http://www.isuma.tv/inuit-knowledge-and-climate-change/movie-no-subtitles.

4

REVISITING THE WILD

Mythology and ecological wisdom in shalan joudry's *Waking Ground*

Leonor María Martínez Serrano

Environmental stewardship

An extremely versatile ecologist working in a wide range of media, Mi'kmaw poet shalan joudry lives with her family in their community of L'sɨtkuk, also known as Bear River First Nation, in Nova Scotia. She has written three books to date: her debut poetry collection *Generations Re-merging* (2014), the play *Elapultiek* (2019), and *Waking Ground* (2020b), which dwells on "Indigenous history, ecological change, and trans-generational trauma" (joudry 2020b, 73). Drawing on Freya Mathews' panpsychism, David Abram's ecophilosophy, and Robert Bringhurst's thinking on wilderness and mythology, this essay explores how joudry's poems in *Waking Ground* are attuned to the power of language to uncover a poetic order in reality, to the legacy and ecological wisdom of her ancestors, and, most important, to a more-than-human world endowed with "agencies of its own" (Bristow 2015, 126). More specifically, this essay ponders the meaning(s) of land for the Mi'kmaq in contradistinction to the epistemologies of control prevalent in the West and argues that wilderness and mythology are of paramount importance in Indigenous worldviews, which advocate empathetic resonance with and respect toward the nonhuman.

Mathews observes that "the way we understand the world, at the deepest level, determines how we live in the world" (2005, 1) and that the current environmental crisis is a crisis of metaphysics. A "revision of the metaphysical premises of our civilization" (1) is thus in order. For this she proposes an antidote, namely, panpsychism (from the Greek *pan* + *psyche*, literally *all mind*), a metaphysic that ascribes mentality not only to humans and nonhumans alike—be they persons, animals, plants, or inanimate natural elements—but to the planet itself as a unity. In other words, panpsychism sees the world as "a subject in its own right with ends and meanings and

DOI: 10.4324/9781003348535-7

communicative capacities of its own and an inclination to communicate with individuals, its own finite emanations" (1). The Earth is, therefore, a kind of "spirit thing," "a One that can communicate with the Many—the 'Ten Thousand Things'—that are its own finite modes" (1). This "spirit thing" Abram calls "a Commonwealth of Breath" (2014, 313), highlighting our engagement with the wild and animate Earth or, as he puts it, "the sensorial reciprocity between our breathing bodies and the bodily terrain" (313). Like other First Nations, the Mi'kmaq are sensitive to the motions of animate Earth, to the common substratum entangling both the human and nonhuman, as attested by their oral literature and mythology. In this regard, *Waking Ground* is peppered with references to a whole mythopoeic *Weltanschauung*, the one joudry inherited from her ancestors with great labor and she now weaves into the fabric of her poems. Put simply, the collection shows her responding to the vitality of the more-than-human world and to the mind of the wild through the lens of Mi'kmaw mythology. After all, a myth is a tool that seeks to shed light on the ultimate nature of reality and uncover the deeper connections that bring humans and nonhumans together.

Waking Ground is punctuated by images of Indigenous trauma and resilience, for joudry is well aware that "centuries of colonialism have resulted in attempts to erase Indigenous presence" through "assimilationist policies and practices such as the *Indian Act*, reserves, and the Residential school system" (M'sɨt No'kmaq et al. 2021, 841). It is also full of images of the land—trees, rivers, bays, birds, fish, butterflies—that remind readers of the need to pay attention to the particulars of the green world, especially in the face of the alarming degradation of the planet brought about by *Homo sapiens* in the Anthropocene. By cultivating a biocentric, environmentally sensitivity attitude, joudry is not just gesturing back to the ecological wisdom[1] and mythopoeic worldview of the Mi'kmaq; she is also countering the mainstream commodification of nature perpetrated by fiercely materialist, neoliberal, capitalist societies, by showing that there is another way of being in the world—one marked by humility, respect, and responsibility. In Indigenous mindsets, "grounded in ethics of care and responsibility for both biodiversity and all peoples" (M'sɨt No'kmaq et al. 2021, 844), "all of the 'ecologies' [are perceived] as alive, interdependent, and interrelating forms in flux and permeated by spirit" (844). A foundational tenet in Indigenous ways of knowing and relating to the land is that "all life forces are sacred and connected" (845), which is one of the commonalities found across the First Nations of North America. Hence it is no wonder that, in the acknowledgments section of *Waking Ground*, joudry should address breathing Earth as a nurturing *oikos* and write: "First, i wish to give thanks to the lands and waters. Without the elements, plant-beings, and animals who give up their lives to feed us we would not be here" (2020b, 73). As if to typographically signal the decentering of humankind and reinforce her ecocentric view of life, joudry insists on writing her name and the first-person personal pronoun "I" in lowercase. As Reynolds (2021) observes,

> [a]s part of her work toward decolonization of language and mindset, shalan joudry chooses not to capitalize when referring to herself in the first person, either as her name or I, to "be consistent with not over-emphasizing myself in relation to the collective."
>
> *(n.p.)*

In joudry's ecocentric view of existence, language reveals itself to be an outgrowth of nature and part of the meaning-making processes at play in the entire world: "it was this maqamikew / who birthed our language" (joudry 2020b, 17), she admits in the poem titled "Here," where she uses the Mi'kmaw word *maqamikew* to designate the *land, ground*, and *landscape* where her community found a home. Grateful as she is "to live close to and have relationships with forests" (73), she is convinced that there are lessons to be gained just by closely listening to nature. Sitting still in the forest, she is "ready to hear the birds and the way the trees sound in the breeze, [...] to just smell the scents of the forest, without naming them or calculating them, just breathing them in" (joudry 2020a). The woods are where she goes to calibrate her mind: "To me, sitting in nature long enough, reminds me about what it means to be part of the land, what it means to be human" (joudry 2020a), she concludes. An advocate of mindful life, joudry writes in "Slow Walking" of how, instead of *knowing* or *taking charge* of the world, she encounters and responds to animate Earth, re-aligning herself with the wild along synergistic rather than impose-and-control lines: walking barefoot on Earth, the lyrical subject feels "terrain / pressing back to you" (joudry 2020b, 57). In sum, the land becomes the ultimate source of meaning, life, knowledge, and identity in the Mi'kmaw mindset.

The meaning(s) of *land*

"Man is altogether too much insisted on. The poet says the proper study of mankind is man. I say study to forget all that—take wider views of the universe. That is the egotism of the race" (Thoreau 1906, 369), writes H.D. Thoreau in a journal entry dated April 2, 1852. In prescient words, the Transcendentalist poet-philosopher anticipates the Anthropocene and laments the troubling impact of the "gigantic institution" (382) of humankind on the Earth, arguing that humans might more properly be a prospect from which to glimpse the infinite. joudry must have closely followed Thoreau's injunction to look beyond humankind and study the nonhuman world. In fact, *Waking Ground* is pervaded by joudry's profound sense of gratitude for the human and nonhuman beings with whom she shares the Earth. It also testifies to the beauty she experiences by closely listening to the more-than-human world.

Simultaneously, these poems seem to imply that our job as human beings is to remember to remember—that is, not to forget the bond uniting *Homo sapiens* with our more-than-human kin and the entire planet. joudry feels herself to be a tiny part in the larger scheme of things, so she does not give up on her mission to listen and respond to an other-than-human world endowed with agency and communicative powers. She does so with reverence, humility, and a sense of deep communion with what-is, such are the intensity and integrity of her attentiveness to our planet. In her encounter with Earth, joudry *breathes through her feet*,[2] aware as she is of the fragility of the biosphere as the home life has built for itself.

Such ecological wisdom is central to the ancestral legacy joudry has been entrusted with as a member of the Mi'kmaq, one of the First Nations to reside in what is now Canada, since time immemorial. Even if her shoulders are not that broad to sustain the legacy of a whole mythopoeic worldview, she feels compelled to perpetuate her ancestors' insights into reality. She does so on behalf of her community—of the no longer living and the still to be born—and on behalf of those who cannot speak to protect themselves in human form, including members of the nonhuman world. Because of the havoc caused by anthropogenic action, she feels that her mission is to compose ecopoems that seek to stir people's conscience into responsible action and reawaken us to our deep bond with the Earth. In fact, owing to "the capacity of environmental texts to model ecocentric thinking" (Buell 1995, 143), poems can act as catalysts for action that quicken environmentalist commitment. joudry's mission is, therefore, not only to compose ecopoems charged with mythological resonance but also to instill in readers a sense of ecological responsibility. She knows that Earth is home, not a pure exterior, as argued by Abram:

> After three and a half centuries spent charting and measuring material nature as though it were a pure exterior, we've at last begun to notice that the world we inhabit [...] is alive. [...] With the other animals [...] we're all implicated within this intimate and curiously infinite world. [...] Inside the world.
>
> *(2010, 158)*

Land is central to Indigenous ways of knowing and being in the world. In a talk titled "What Does the Earth Ask of Us?," professor Robin Wall Kimmerer addresses the question of what the term *land* means in Indigenous worldviews. In our Western mindset, she argues, the land is understood first and foremost as property, capital, and commodity; that is, as "a source of ecosystem services (e.g., plants making us oxygen to breathe, water purification, etc.)" (Kimmerer 2021). In contrast, in many Indigenous peoples' worldviews, the land is seen "not as property for which humans would claim some rights, but as something for which we have deeply embedded moral responsibility" (Kimmerer 2021). The twin concepts of *land* and *identity* are inseparable in the Indigenous *Weltanschauung*, where the land is perceived as being a sustainer that takes care of us and as the residence of people's ancestors and nonhuman relatives. Land is conceived as a healer, a pharmacy, a caretaker. Land is also understood as a library, a source of knowledge, and a teacher. It is sacred and literally "enspirited"; that is, it is populated by spirit beings and the locus of a profound connection to ancestors and descendants. Put simply, land is home and a gift. The natural laws of Indigenous peoples acknowledge that "the very essence and source of life—water, air, and soil—is the right of every living thing," that "man and nature are one," that "everything comes from the land," and that "all that the earth holds is sacred" (M'sɨt No'kmaq et al. 2021, 846). In light of the importance of land in Indigenous worldviews, it is no wonder joudry should ponder the meaning(s) of *land* in several poems that are central to the overall architecture of *Waking Ground*. Land is already implicit in the very title, which gestures toward an animistic conception of the universe typical of preliterate, preindustrial societies, including the Mi'kmaq. In the Mi'kmaw mindset, the land is a living being and is attributed communicative powers and agency. In this respect, Abram has postulated the notion of "Commonwealth of Breath" (2014, 313) to refer to the Earth as being *a whole of wholes* that brings together human and nonhuman beings into a totality of material–semiotic interconnections. The world where humans dwell is "our larger flesh, a densely intertwined and improvisational tissue of experience" (Abram 2010, 143) and all bodily natures—all embodied, *enworlded* beings—are kindred, "interdependent constituents of a common biosphere, each of us experiencing it from our own angle" (143). In terms reminiscent of Mathews, to Abram's mind the Earth is an indivisible (infinite) unity (the One), which self-realizes and manifests itself in the (finite) Many populating the Earth.

In *Waking Ground*, the concept of *land* is looked at from multiple perspectives in joudry's attempt to shed light on what it means for humans and nonhumans to be earthbound beings. Thus, in a poem titled "Here," the poet dwells on *land* as the ultimate source of nourishment and meaning for human beings. "Here" has the specificity of a context-bound deictic and, as such, refers to the place that is closest at hand at the moment of enunciation. Yet it also conveys a sense of universality, so that *here* is *anywhere on Earth*. Writing on the origin of language, joudry says: "It was this maqamikew / who birthed our language" (2020b, 17), deliberately using the Mi'kmaw word *maqamikew*, meaning "land; ground, landscape" (17), to stress her connection with her homeland. In the Mi'kmaw worldview, language and meaning ultimately come from the land, in much the same way that trees, mountains, and rivers ultimately owe their existence to the same telluric principle. As Abram notes, language is "rooted in the non-verbal exchange always already going on between our own flesh and the flesh of the world" (1996, 90). More important, among oral peoples, language serves a wide range of functions other than mimesis, as it "functions not simply to dialogue with other humans but also to converse with the more-than-human cosmos, [...] to invoke kinship even with those entities which, to the civilized mind, are utterly insentient and inert" (Abram 1996, 70–71). If the Earth is the universal *oikos*, then it is only natural to think that the land is the origin of everything that exists on this planet, including the words we use to communicate with each other.

As Gary Snyder observes in "Language Goes Two Ways," there is something essentially wild about language: languages are complex living organisms and "naturally evolved wild systems" (2000, 127). In "Tawny Grammar," Snyder dwells on how there is room for wilderness even within human languages themselves: "The grammar not only of language, but of culture and civilization itself, is of the same order as this mossy little forest creek, this desert cobble" (1990b, 76). Wildness is thus ubiquitous, a property of Earth, *Homo sapiens*, culture, and language. Along similar lines, Abram contends that language is "born of the interplay and contact between the human and the more-than-human world" (2010, 95). In speaking Mi'kmaw, joudry's people are reminded of "the way rivers move" (joudry 2020b, 17), for language ultimately mimics the motions intrinsic to the world. In this respect, the poem titled "Poems and Streams" explores the isomorphisms between language and Earth through illuminating metaphors. joudry writes lines with the texture of simplicity and transcendence: "these words are water / these lines are moments" (18), which is meant to suggest that her poems are made of the stuff everything else is made of. As such, they are also part of the nonstop flux that characterizes reality and evokes Herakleitos's notion that nothing ever stays quite the same. In joudry's view, the words in a poem recall water, one of the ultimate sources of life identified by the pre-Socratic philosophers in the fifth century BCE. Upon closer scrutiny, the Mi'kmaw mythopoeic worldview is not completely dissimilar to that of the poet-naturalists of ancient Greece, whose aim was not to establish dominion over reality but to comprehend the subtle ways in which everything was connected to everything else by a common material–spiritual substratum. In mimicking or trying to capture the constant metamorphosis of the more-than-human world, joudry's lines become "unpunctuated / flow" (18), evoking the poetic order implicit in the cosmos and resonating with the energy circulating all around. joudry's poetry is thus the song of the flesh of the Earth; it responds to the calling of the planet by becoming breath and music beating to the rhythm of what-is.

Meaning is not the sole prerogative of *Homo sapiens*; rather, the world is communicative and the ultimate source of meaning. Abram has come up with illuminating insights into the communicative powers of the Earth and the meaning-making processes at work in the more-than-human world. The eco-philosopher conceives of language as being "a property of animate earth" (Abram 2010, 11), not an exclusively human possession, but rather an attribute of all beings, sentient and nonsentient alike. All things and beings have thus "the ability to communicate something of themselves to other beings" (Abram 2010, 172). As such, human speech arose as a response to an expressive "cosmos that already *spoke to us* in a myriad of tongues" (4), and so its fundamental function is not mimesis "but to call ourselves into the vital presence of that world—and into deep and attentive *presence* with one another" (11). An act of attention and presence is precisely what is at stake in the poem "Tupkwanamuksi" ("Brown(ing)"), offered in both Mi'kmaw and English, where readers witness the lyrical subject's communion with Earth. Lying "on bare landscape" (joudry 2020b, 19), the poetical self is seen embracing the Earth as though it were her lover and caretaker. Once again, joudry refuses to capitalize the first-person personal pronoun: "i hold you dearly (reluctant to part)" (19). Earth is depicted as being not inert matter or sheer exteriority but as a lover giving warmth and protection instead.

In joudry's conception, Earth is also a compass helping humans not to get lost, as suggested by two poems of extreme brevity, "How to Not Get Lost" and "Train Tracks and Pipelines." In a world marked by a profusion of human-made objects and "manufactured things" (joudry 2020b, 63), nature remains the guiding compass that helps us navigate space safely. In the former poem, joudry writes that the best course of action for people to take so as not to get lost is to "follow a line toward the tallest tree" (24) and from there to keep on following the line to the sea, which has been an important presence in the Mi'kmaq's lives for millennia. In the latter poem, by contrast, the piece of advice given is to use "manufactured things" (63) as a path leading to town, replacing the sea from the former poem with a cityscape as the framework of reference in industrialized societies. In this connection, because of the proliferation of human-made things, it has become increasingly hard to draw a clear-cut boundary between the natural and the artifactual. Anthropogenic actions have changed the face of the Earth in dramatic ways, particularly since the period of the Industrial Revolution. In fact, given the growing "impacts of human activities on earth and atmosphere" (Crutzen and Stoermer 2000, 17), Paul J. Crutzen and Eugene F. Stoermer predict that "mankind will remain a major geological force for many millennia, maybe millions of years, to come" (18). "Train Tracks and Pipelines" gestures thus to the profound changes of the Earth brought about by anthropogenic action.

The focus is on bareness in "Slow Walking," a poem prefaced by a textual threshold attributed to Mi'kmaw Chief Frank Meuse that is an apt reminder of the widespread lack of attention to and alienation from our physical environment. In this regard, joudry embraces a mindful attitude in our way of being in the world and proposes that one go to the wilderness to calibrate one's mind, which is but a prolongation of the mind of the wild. The poet is not alone in imputing a psychic dimension to the biosphere, though. In Abram's ecophilosophy, consciousness is not "an interior human trait" that comes into full bloom in our skulls but rather "a ubiquitous quality of the world" (2010, 37). Along the same lines, Bringhurst argues that "[t]he land has a mind of its own" (2018, 9), acknowledging that the Earth is endowed with a vast consciousness and that *Homo sapiens* is not the sole possessor of mind. For her part, Mathews has called this conception of the world *panpsychism*, which she defines as *any* view that "attributes a psychic dimension to all physicality" (2003, 4), dismantling thus "the foundational dualism of Western thought" (4). Drawing on Leibniz's and Spinoza's "intuition that everything is woven from the same skein, that everything informs everything else, and hence that 'mind' in some sense must be intermingled with 'matter'" (Mathews 2003, 3), and drawing on their "intimation of interpermeation—of everything participating in everything else, of mutual holding and enfolding at the structural base of existence" (3), Mathews embraces a view of the world that "implicates mind in matter and matter in mind" (27) and posits a form of "nonduality or mind-matter unity" (27) that transcends the Cartesian dualistic conception of mind and matter as "categorially distinct substances" (26). Hence, Mathews views the world in nondualistic terms, as communicative subject. Humankind's aim should not be to *know* but to *encounter* the cosmos, which is understood as "an indivisible unity" (Mathews 2003, 47), "a unified, though internally differentiated and dynamic, expanding plenum" (47)—a self-realizing, self-actualizing, self-differentiating unity. The universe consists of "(finite) self-organizing systems, or selves, that constitute centers of subjectivity in their own right [...] [and] may be described as the [finite] Many into which the primordial subject, the [infinite] One, differentiates itself" (Mathews 2003, 55). If the world is a subject endowed with a mind of its own, then the right way to relate to it is not the pursuit of *knowledge* but the experience of *encounter*, argues Mathews.

In its brevity and gnomic conciseness, "Slow Walking" is instructive in multiple respects as a poetic enunciation of Mathews' panpsychism. joudry seeks to immerse her *self* in the world, "to participate fully in the realness of the world" (Mathews 2003, 58), not to establish dominion over Earth. By way of evocation and indirect statement, "Slow Walking" reminds readers that feet are our primordial link with the land, ground, or earth. Having one's feet firmly planted on the earth is expressive of the fact that humankind belongs in a wider, sensuous totality of (non)human sentient beings. Walking barefoot somehow enacts and responds to the ancestral call of the terrain, "pressing back to you" (joudry 2020b, 57), affirming time and again the kinship uniting all beings. In joudry's memorable words, each step is "a call and return / of wild things" (57), stirring memories of a time when humankind felt at home on Earth rather than alienated from a deanimated world. This happened to be the case with the Mi'kmaw and other Indigenous people of North America. It was not the aspiration of the First Nations living in preliterate, preindustrial societies to "run the world or tame it" (Bringhurst 2018, 8). In fact, their oral literatures provide ample evidence that they remembered their ancestral link to Mother Earth: "[t]heir stories tell us that they understood that the wild is in control of itself and has room within it for humans but does not need and cannot tolerate human domination" (Bringhurst 2018, 8–9). "Slow Walking" celebrates the ancestral bond and universal kinship between the human and the nonhuman, while advocating a form of dwelling in this "Commonwealth of Breath" that is respectful to all earthbound creatures.

The wilderness as library

In a journal entry dated March 15, 1852, Thoreau celebrates the arrival of spring with the pronouncement of his desire to do better and live fuller, "to have my immortality now" (1906, 351). He writes: "My life partakes of infinity. The air is as deep as our natures" (350) and "I am eager to report the glory of the universe; may I be worthy to do it" (351). The next day, March 16, he spends the whole day in a completely different kind of library, Cambridge Library. Surrounded by oceans of volumes, he comes up with the notion that the library is "a wilderness of books" (352). However, not all books are true books; only a few are needed for readers to attain a grain of truth. With unstudied simplicity, he writes:

> Books which are books are all that you want, and there are but half a dozen in any thousand. [...] [W]e are accumulating a forest of books in our rear, as wild and unexplored as any of nature's primitive wildernesses. *(353)*

Thoreau is juxtaposing here two kinds of libraries and two ways to encounter and immerse oneself fully in the world. To American Transcendentalists, knowledge was to be gained through great labor, and not just indoors, in the silent company of books and in the warmth of a traditional library crowded with ancestors' precious insights. They also read and studied the lines inscribed in the book of Nature and understood that the wilderness was a vast library written in nonhuman characters. In the company of trees and stones, rivers and fungi, grass, and birds, it was possible to attain ecological literacy and experience learning of lasting value. As Bringhurst puts it, "Sun, moon, mountains and rivers are the writing of being, the literature of what-is. Long before our species was born, the books had been written. The library was here before we were. We live in it" (2006a, 143). Like Thoreau and her Mi'kmaw ancestors, joudry actively seeks the company of the forest. It is only natural that some of the poems central to *Waking Ground* should explicitly address arboreal existence and capture the wisdom of trees: "Rock Saplings," "Raising Forests," "Grounded," "Rebirth," "Siwkw," and "Apuknajit" are among them. They show joudry conversing with the more-than-human world, listening with her eyes, interpreting trees and their place in the larger mesh of living things—in a world where everything is related to everything else and shares a common spiritual–material substratum. She does the kind of reading Bringhurst describes in his prose meditation "Reading What Is":

> We were reading waves and rivers, winds and clouds, the tracks of moose and grouse and hare, long before we started reading words. The reading we do now [...] owes its life to that apprenticeship in *paying ecological attention.*
>
> (*2006b, 195*)

In "Rock Saplings," joudry dwells on a universe of elemental things—rocks and trees—and on the eternal cycle of regeneration intrinsic to nature. The poet chooses to focus on the minutest details of the forest, as exemplified by rock saplings, that is, small trees *in potentia* "birthed on a boulder" (joudry 2020b, 20), trying to stretch their boughs upwards in search of sunlight to survive. Though they will never "reach the company of canopy" or "taste the air moved by a hawk's wing" (20), they "keep vision skyward" (20) on account of their resilience and will to thrive. The lyrical subject tends to them, aware of their vulnerability and their key role in "the recomposition of elements" (20), for by returning to soil they give "life to other small things" (20) in a never-ending cycle of death and resurrection. The vibrant matter out of which the whole Earth is made is a substratum common to all dwellers on this planet. The rock saplings are born and die with a purpose within the larger scheme of wild things. But what exactly is the meaning of *wild*, of which the forest appears to be the perfect embodiment in joudry's poems in *Waking Ground*? As Bringhurst notes in "Wild Language," *wild* is not synonymous with "undisciplined, unpredictable, savage, frightening, fierce, raw, crude" or "formless" (2006c, 261). Rather, *wild* means "undomesticated, unmanaged, uncontrolled by human beings," "*extremely sophisticated*" and "self-sufficient" (261, 262). In Bringhurst's conception, the wild is the true benchmark and standard by which to measure our actions on Earth. It is not "something to conquer or subdue; it's something to try to live up to" (268). It is also the ultimate source of meaning and knowledge: "[t]he wild is the real, and the real is where we go for form and meaning. Meaning doesn't originate with us" (264). Bringhurst observes that "[t]he wild is the biosphere" (269) and *Homo sapiens* is "so deeply enmeshed in the fabric of nature that all separation is an illusion" (266). The wilderness has room within it for our species, but it does not need anything from us and "cannot tolerate human domination" (Bringhurst 2018, 10). Given our interdependence with the nonhuman world, Bringhurst suggests that we rethink how we relate to the Earth: "The earth's life is much larger than our own lives, but our lives are part of it. If we take that life, we take our own" (12). For his part, in "The Etiquette of Freedom," Snyder observes that wildness is pervasive and ubiquitous, "it is everywhere: ineradicable populations of fungi, moss, mold, yeasts, and such that surround and inhabit us" (1990a, 14). *Homo sapiens*' embodied minds are also wild: "the depths of mind, the unconscious, are our inner wilderness areas" (16). So is language.

In "Raising Forests," joudry mourns the disappearance of trees from the face of the Earth. Elegiac in tone, the opening line of this poem testifies to the endless havoc caused by our species on the planet and its (non)human dwellers and, more specifically, on the homeland of joudry's ancestors: "we are not the first generation to lose forests / as trees were slain for ships and forts" (2020b, 50), she writes, evoking centuries of colonialism in Mi'kmaw territory. From an arboreal perspective, the history of humankind is the relentless succession of "massacres of forests" (50), which were parallel in time to the damage inflicted upon whole communities of Indigenous people in North America by the white man's colonizing endeavor. Whole communities were deprived of their traditional territory and their "livelihoods and continuity" (50) were fractured, as their lands were cleared and their forests were cut down to build new ships and forts. Forests and trees deserve better than that, though. In the closing lines of the poem, joudry's voice calls for empathy and collective action to protect forests and the environment at large: "let's tend to the forests like prophets / encourage them to wilder in old growth / and watch them mature into being" (50). These lines convey a moving eulogy of trees at a time when "the dominant human culture is increasingly toxic to the wild" (Bringhurst 2018, 20). Faced with humankind's determination to destroy Earth and all forms of life in a self-suicidal race, joudry encourages us to tend to the forest, to participate in the biosphere with empathy and responsibility. In the Anthropocene, at a time of alarming environmental degradation, humility and attention to the more-than-human world are badly needed. Like other ecopoets on the Canadian scene—Bringhurst, Jan Zwicky, Don MacKay, and Tim Lilburn, among others—joudry invites readers to spend time in the wild, embracing a form of reverential silence that leaves all egotism behind. This form of listening does not seek to impose or control, but to experience the world with humility and gratitude, as "*we don't own what we know*" (Bringhurst 2006c, 270).

In "Apuknajit" (the Mi'kmaw word for *February*), the lyrical subject finds a true haven in the forest. Looking out her window at a snowy landscape, tea "brewing / on the wood stove" (joudry 2020b, 45), she meditates upon a frozen world and suddenly comes to the realization that the woods are a space for healing one's mind: "i know how a cure for many things / is to walk out into it" (45). The forest is also a space for epiphany and sudden illumination, as suggested by joudry's gentle invitation to "observe the stillness in small things / tame the mind's bustling chatter" (45), expressing thus the essence of mindfulness. The valuable lesson she delivers to those willing to listen is that, by looking intently at the stillness present in the forest, we might learn to remain attentive and, in doing so, still our very minds. By realigning our mind with the mind of the wild in reverential silence, we might achieve peace and quiet. In an ecocentric and biocentric mindset like the one joudry embraces, the forest is not owned or controlled but rather experienced to the fullest degree.

In the same spirit, the poem "Grounded" celebrates the entanglement of the human and the nonhuman. The lyrical voice refers to the forest in terms that are expressive of love: "i lay in forest's arms" (joudry 2020b, 67). Fully enmeshed in nature, cradled by ancient trees, the lyrical "I" experiences a process of complete immersion in the scents and views of the woods. Like a vast Emersonian eyeball, she perceives the specifics that are unique to the place: the green canopy hanging over her head, the "fungal scent" (67) of mosses, and the "night's vestigial rain leaking / from fractal branch and sky" (67). Nothing escapes her enhanced attentiveness, honed by constant exposure to the elements making up the forest. The epiphany comes when the lyrical voice reveals that she is a small part *of*—not apart *from*—this very world she seeks to paint in words. "[C]oloured by humus and stone" (67), feeling very small in the larger scale of nature, she senses, like other nonhuman beings (slugs, deer), the ancestral call of the ground. Ultimately, this act of sensuous encounter with the forest and its dwellers is expressive of a shared becoming of perceiver and perceived, or a dynamic of mutual arising where Mathews' infinite One materializes itself into the finite Many—that is, the embodied, ensouled, earthbound creatures that form an indivisible unity.

Even if eco-elegy appears to be pervasive in *Waking Ground*, there is room for hope in joudry's collection, too. In "Siwkw," the poet sings of the arrival of spring in terms reminiscent of Thoreau's 1852 journal entry quoted above. Seriously damaged trees are able to mend themselves and rise again. Looking at them "leafing-out again" (joudry 2020b, 69), the lyrical subject is astonished to see how fatal wounds in trees heal and they thrive in splendor with the resurrection of the natural world. Once again, the self perceives itself to be "a small piece of the land / holding on" (69), persisting in time despite the havoc caused by humankind. There is also space for hope in the poem titled "Rebirth," where trees speak in the first person. Contemplating "the fleshy world below" (70) from the heights of their uppermost boughs, trees mourn "a whole generation / missing" (70) on account of relentless deforestation at humans' hands, yet they are amazed at their own resilience and their capacity for self-regeneration. They keep on bursting "from the soil / in full bloom" (70), "learning the art of the waking ground" (70). It is no coincidence that "Rebirth" should be the closing poem in *Waking Ground* and that the closing words of the poem and book should be the ones that give the collection its title. The ground or land is numb and frozen in wintertime, awaiting the arrival of spring with its promise of renewed life. In springtime, the ground wakes up to call all earthbound beings back to life, to remind them of their ancestral roots and of the imperative to persist despite the piecemeal destruction of the world brought about by dominant industrialized societies. The waking ground is the common link or substratum that brings together trees, the "thin strands of fungi" (70), the "earth's own bones" (70), and "the hardest kuntul" (*kuntul* is the Mi'kmaw word for *rocks*) into an indivisible whole. They are all part of a gigantic material–semiotic network of interconnected beings.

Language, myth, and Indigenous identity

Language is an inextricable part of people's identity and "reflects ways of knowing" (M'sɨt No'kmaq et al. 2021, 850). In the Mi'kmaw worldview, "all stories, learning, and language come from the land" (840). What is more, languages are derived "from the sounds and rhythms of ecology, nature in action" and constitute "repositories of science, they tell of relationships, they reveal history, and they hold Indigenous identity" (846). According to the 2016 census, 8,870 people (out of the 58,763 people registered with Mi'kmaw First Nations as of 2015) were listed as speaking the Mi'kmaw language (Gallant 2021). However, according to a 1999 report by the Nova Scotia Mi'kmaw Language Centre of Excellence, there were fewer than 3,000 fluent speakers (Gallant 2021). joudry is among the roughly 110 people who live in a rural community in L'sɨtkuk and are deeply concerned about the future of Mi'kma'ki. The poet speaks with reverence of her ancestors' language, a vast treasure-house of the knowledge of the land. She believes that learning Mi'kma'ki is like opening a door into the past and stepping into a different worldview; it is tantamount to experiencing the world through her ancestors' eyes. Acknowledging the deep bond between her native language and the land, she speaks thus in an article published by Canadian Broadcasting Corporation (CBC):

> Our ancient Mi'kmaw ancestors were the first humans on this landscape and the ecosystems helped shape the people's culture. Our elders [...] tell us that we sprouted here, *weji-sqalia'tiek*, just as any plant comes into the world from a seed full of history and possibility. [...] [T]he Mi'kmaw language holds the rest of the story of the plants, animals and geography as it can describe the land and the changes over time.
>
> *(CBC 2019)*

Having grown up in a home where Mi'kma'ki was not spoken, joudry feels ashamed that she cannot speak her people's language. It was not her choice as a child to decide to learn or not learn the language, because she was not given opportunities to learn it in public schools, and "there were no films, radio programs or books in the language in my young adult years" (CBC 2019). Staring down the precipice of language loss, joudry mourns the disappearance of Indigenous languages, including Mi'kma'ki. As generations of fluent speakers are slowly passing away, it seems an impossible mission to create new fluent speakers and ensure the continuity of Mi'kma'ki. Yet joudry feels it is a moral imperative for her "to rescue the language from the fluent individuals in other regions and pass it on to our children" (CBC 2019). She elaborates on her fight to keep her language alive: "reclaiming and restoring our language in families or regions with few fluent speakers feels like a daunting responsibility. [...] We're fighting to save our language and it's a race against time I'm not sure we will win" (CBC 2019).

As joudry points out, the 1876 *Indian Act* removed "Indigenous culture from the people. One of the mechanisms to do this was to remove Indigenous children from their homes and place them in residential schools where they were to only speak English and practice Euro-Canadian customs" (CBC 2019). The *Indian Act* left an indelible mark on generations of Mi'kmaq, as it sought to assimilate a vast spectrum of First Nations into mainstream, non-Indigenous society. It forbade Indigenous peoples "from expressing their identities through governance and culture" (Parrott 2020), which led to the termination of their cultural, social, economic, and political distinctiveness. In other words, the *Indian Act* allowed the government to control most aspects of Indigenous life, ranging from land and resources management, through Indian status and band administration to education, all of which resulted in long-term poverty and marginalization, "trauma, human rights violations and social and cultural disruption for generations of Indigenous peoples" (Parrott 2020). One of the dark legacies of this system was that "people coming out of the schools could not or would not speak their native language, therefore they did not pass it on back home and when they raised their own children" (CBC 2019), which accounts for the dwindling numbers of fluent Mi'kma'ki speakers. joudry admits her wish to learn Mi'kma'ki and be part of her community's healing process. For over a decade now, she has taken "long trips to other communities and regions where the Mi'kmaw language is more prevalent" (CBC 2019) to learn from fluent individuals, despite dialectal differences typical of a "still primarily oral-based language" (CBC 2019). She is aware that "[i]f the language becomes extinct, the knowledge within it also becomes extinct, including knowledge of relating to the land" (M'sɨt No'kmaq et al. 2021, 849). As home-schooled Cedar Meuse-Waterman puts it, "The language is like a doorway which leads to the true meaning of things" (CBC 2019). joudry and her daughters practice a few words and phrases of Mi'kmaw together every day to keep the language alive. However, bringing "fluency back to families" cannot be sustained by individual responsibility alone, but needs to be supported by "language programming in our schools and communities" to ensure that everyone is "on a path of healing" (CBC 2019) the wounds from the past. A bilingual notice in Mi'kma'ki and in English welcoming visitors to joudry's community's reserve, established in 1820, reads thus: "We, the people of the Bear River First Nation, will work towards healing our minds, bodies, spirits and environment" (CBC 2019). Human and environmental justice go hand in hand.

It is against this backdrop that joudry's fight to preserve her ancestors' legacy is to be understood. In some poems included in *Waking Ground*, joudry denounces the traumatic experiences undergone by the Mi'kmaq for centuries, under the forces of colonialism. "Digits" is a case in point, inasmuch as it is a fierce denunciation of the common practice among colonial regimes to subjugate and acculturate Indigenous peoples in North America. Deprived of their territories and livelihoods, of their languages and cultures, forced to live within "invisible fences" (joudry 2020b, 46), people's "minds were able to shed their skins" (46), to anesthetize themselves in the face of the singularity of their ancestral roots. The Mi'kmaq were given numbers to mark their identity, to make them fit in a new scheme of things and get them ready for "public consumption" (46). joudry writes with nostalgia about how her ancestors got to naturalize the rules devised by a perverse system to control their identity, "as if being l'nu was a thing of numbers" (46). The poet uses here "l'nu," the Mi'kmaw word for "Native person or Mi'kmaw (singular)" (46). This is still part of the unresolved legacy of colonialism in Canada.

Mythology plays a crucial role in the Mi'kmaw worldview. According to Bringhurst, a myth is "a theorem about the nature of reality, expressed not in algebraic symbols or inanimate abstractions but in animate narrative form" (2007, 63). A mythology is "an ecosystem of myths" (63) that embodies a complex worldview, one that acknowledges the presence of spirit beings as part of the very fabric of the more-than-human world and is marked by humans' respect for and gratitude to the nonhuman. Myths are therefore not to be dismissed as naïve explanations of the world, typical of preliterate, preindustrial societies that conceived of the world as being animate. On the contrary, Bringhurst argues that "[m]yth is an alternative *kind* of science, [...] a means of understanding and elucidating the nature of the world" (64). Like other First Nations, the Mi'kmaq tell creation myths to account for the origin and endless fascinations of the world. In Mi'kmaw oral tradition, the world and all its inhabitants were created in seven stages. The Great Spirit or "Creator made the sky, the sun, Mother Earth and then the first human beings, i.e., Glooscap and his grandmother, nephew and mother" (Gallant 2021). Glooscap commanded seven men and seven women to come forth from sparks of fire; they would become the founding families of the seven Mi'kmaw districts in the traditional territory in the coastal areas of Gaspé and the Maritime Provinces east of the Saint John River, which the Mi'kmaq have occupied since time immemorial (Gallant 2021). This land has sustained the Mi'kmaq for more than 10,000 years and so they have developed a strong reverence for their environment. "Seasonally patterned habitation and resource harvesting" made them spend "spring and summer on the coast, fall and winter inland," using shellfish, sea and land mammals "for nutrition, clothing, dwellings, and tools" (Gallant 2021). It is no wonder that the Mi'kmaq believe that a good life amounts to living in harmony with the more-than-human world and its (non)human dwellers.

For the peoples of the First Nations, "[s]tories are the platform for teaching Indigenous worldviews, values, culture, and how to live with and uphold responsibilities to the land" (M'sɨt No'kmaq et al. 2021, 847). Stories are thus powerful vessels of meaning in their oral literatures, a way to preserve fundamental life lessons and pass on valuable knowledge to the younger generations. As evidenced by the poem "Wiklatmu'jk" (the Mi'kmaw word for "little people of the forest"; joudry 2020b, 47), joudry practices the art of storytelling and story listening with her own daughters to preserve and promote her ancestral legacy. She is aware that her stories were once "the birthing space" (47) whereby she voiced "improvised teachings / meticulous but careful" (47) to pass on to her daughters. With age, she comes to the realization of "how empty the forest is without *the people* / how sterile our lives" (47), with "the earth's tricksters misplaced" (47), and she mourns the "little magic" (47) left in her. Yet she witnesses the joy with which her daughters are still participant in a world animated by spirit beings, moved by the fact that "mythology's last umbilical cord" (47) is still alive in them. When her daughters asked for more stories and she remained silent, she would "take them to the trees" and tell them to look for *wiklatmu'jk*, disguised under "*a different shape*" (47), for in the Mi'kmaw worldview the cosmos is pervaded by spirit beings and form-changing tricksters. In light of the unresolved legacy of Canada's settlement, myth and storytelling as practiced by contemporary poets like joudry constitute a most effective tool to express truths of durable value and heal the wounds resulting from centuries of colonialism.

Myth and history intertwine in *E'sekati*, a complex poem where joudry explores 11,000 years of shared history of the land and the Mi'kmaq, as well as the trauma experienced by her ancestors under the conditions imposed by colonialism. The result is an evocative poem of powerful mythical resonance and the beating heart of a documentary film by Bretten Hannam and joudry released in 2017. In capturing the uniqueness of the land and the multilayered history inscribed in the terrain, the poet grieves her people, "disentangled from landscape to make way / forced to make way" (joudry 2020b, 55) after their first contact with colonizers. joudry offers thus a general overview of the Mi'kmaq's colonial history after their first contacts with adventurers and settlers. Owing to their geographical proximity to the Atlantic, the Mi'kmaq were "among the first peoples in North America to interact with European explorers, fishermen and traders" (Gallant 2021), which led to an alarming decimation of their population and sociocultural disruption. The havoc caused by the first sustained European settlements in what is now Canada was beyond words. Exposure to smallpox and other European diseases to which the Mi'kmaq were not immune resulted in a "loss of up to half the population from about 1500 to 1600" (Gallant 2021), and their lifestyle was also dramatically altered, shifting from subsistence hunting and gathering to trapping and fur trade (Gallant 2021).

The Mi'kmaq's history is punctuated by treaties with colonizing forces whereby they sought to protect their rights to the lands they had inhabited for as long as they could remember. They were "largely allied with French colonial forces," which had permanent "settlements across Acadia until the 18th century" (Gallant 2021) and sustained prolonged conflict with the British. After conflicts with Britain, the Mi'kmaq also signed up to four different treaties (known as the Peace and Friendship Treaties) in the 1700s and two more treaties during the American Revolution to secure alliances and have their rights to the lands recognized (Gallant 2021). The struggle continued in the 19th and 20th centuries. Life under British, and later Canadian, governance inflicted irreversible damage upon the Mi'kmaq and their lifestyle. Attempts to "establish them as agriculturalists failed because of badly conceived programs" (Gallant 2021) and the Mi'kmaq ended up joining the workforce of the 19th- and 20th-century economy as craftsmen, fishers, road, rail, and lumber workers. Worse was "the lasting trauma of residential schools" and the relocation of "more than 2,000 Mi'kmaq people living in numerous small communities to government-designated reserves" (Gallant 2021) in the 1940s. Deprived of their lifestyle and living in reserve lands, "facing discrimination" and "a lack of civil rights" (Gallant 2021), generations of Mi'kmaq have experienced "centuries of systemic prejudice and persecution rooted in colonial worldviews and practices" (M'sɨt No'kmaq et al. 2021, 842). The wounds still need to be healed.

Invoking her *ancient people* (*sa'qewe'kl'nu'k), not so recent people* (*mu awsami kejikawe'k l'nu'k*), and *recent ancestors* (*kejikawe'k l'nu'k*), joudry retells the Mi'kmaq's history in "E'se'kati," the Mi'kmaw word for a town called Lunenburg, located in their traditional territory. Eleven thousand years of history in "the land of dawn" (*wapne'kati*)—that is, the "larger Indigenous region of Eastern Canada, which includes the Mi'kmaq territory" (joudry 2020b, 54)—are captured with astonishing linguistic economy in the three opening stanzas of the poem. "Chasing giants across territory" (54) and relying on their surroundings, joudry's ancestors found themselves a home in an inhospitable landscape, on "such unrelenting coast" (54), "the harbour of clams" (54). They made use of "bowls of clay" (54) that would then return to the ground, they welcomed the sun every day, gave blessings for everything Mother Earth gave them for sustenance, and looked after the land with profound gratitude. Contact with European settlers unleashes "the bringing and taking of lives" (55) that results in dramatic population decimation. A litany of the woes and havoc caused by colonialism ensues: "bury those hearts in your harbour of clams / between the calligraphy of stunted trees" (55). A long "geology of families" (55) is inscribed in the terrain and blood spilled on the ground by subsequent colonizers: "the blood of french who dug farms / the blood of english who dug lines / the blood of boat people who were loyal to the crown" (56). Towns were built and treaties were signed in the meantime to protect the Mi'kmaq's inherent rights to their traditional territory and resources. In the closing lines of "E'se'kati," joudry begins the path toward healing, hers and her people's, with her heart immersed in the rhythm of the ocean's waves, "inhaling / exhaling / lifting debris" (56).

In joudry's ecocentric view of existence and human history there is only one way out, which is to heal the wounds from the past and "reconcile the distance between us" (joudry 2020b, 68), by embracing humility and respect toward the more-than-human world. In the mythopoeic poem titled "Tending the Fire," she invites readers to gather around a fire and listen to her ancestors' stories so that we get to "understand the shapes of what was / and weep suddenly with me" (68). What has vanished is a lifestyle in communion with the more-than-human world where knowing was not synonymous with owning and where myths formulated theorems about the nature of reality. Like poetry, science, and philosophy, mythology is also a form of knowing. Under the pressing forces of colonialism, the Mik'maq's language and worldview were on the verge of irreversible extinction. Her hope is that it is not too late to preserve her ancestors' legacy, part of which is the ecological wisdom that *Homo sapiens* is not the center of the cosmos. Finding in nature a true compass to human existence, she writes: "let us belong to the land, instead / know the magnetism of trees / watch the sun find its farthest peak" (68), without calling them hers or ours. Ultimately, in its deft denunciation of injustice—toward human communities and the more-than-human world—*Waking Ground* encapsulates a wealth of wisdom that reveals itself to be pertinent in the Anthropocene. By opening readers' eyes through the medium of mythology to ways of dwelling on and relating to Earth with a sense of duty and responsibility, joudry's poems teach ethical lessons of lasting value: that humans are not the measuring rod of the planet, that the Earth is a Commonwealth of Breath where all beings are deeply interconnected and interdependent, and that all forms of life are sacred and deserve protection and respect.

Notes

1 The Mi'kmaw concept of *M'sɨt No'kmaq*, meaning "all my relations," gestures to "a kin-relationship with the land, waters and all living beings" (M'sɨt No'kmaq et al. 2021, 840). In fact, "[t]here is growing recognition of the effective governance of Indigenous lands by Indigenous Peoples, with 40% of the Earth's most intact remaining biodiverse areas in Indigenous stewardship" (843).

2 This metaphor is borrowed from Bringhurst's essay "Breathing through the Feet: An Autobiographical Meditation" (1986), where he contends that hands are more damaging than feet in our commerce with the Earth and hence the injunction "breathe through your feet" is a better option than "pay attention."

References

Abram, David. 1996. *The Spell of the Sensuous. Perception and Language in a More-than-Human World*. New York: Pantheon.

Abram, David. 2010. *Becoming Animal: An Earthly Cosmology*. New York: Knopf Doubleday.

Abram, David. 2014. "Afterword: The Commonwealth of Breath." In *Material Ecocriticism*, edited by Serenella Iovino and Serpil Oppermann, 301–314. Bloomington and Indianapolis: Indiana University Press.

Bringhurst, Robert. 1986. "Breathing through the Feet: An Autobiographical Meditation." In *Pieces of Map, Pieces of Music*, 99–110. Toronto: McClelland & Steward.

Bringhurst, Robert. 2006a. "Poetry and Thinking." In *The Tree of Meaning: Thirteen Talks*, 139–158. Kentville, NS: Gaspereau Press.

Bringhurst, Robert. 2006b. "Reading What Is." In *Reading Writers Reading: Canadian Authors' Reflections*, edited by Dannielle Schaub, 195. Edmonton: University of Alberta Press/Jerusalem: Hebrew University Magnes Press.

Bringhurst, Robert. 2006c. "Wild Language." In *The Tree of Meaning: Thirteen Talks*, 257–276. Kentville, NS: Gaspereau Press.

Bringhurst, Robert. 2007. "The Meaning of Mythology." In *Everywhere Being Is Dancing: Twenty Pieces of Thinking*, 63–72. Kentville, NS: Gaspereau Press.

Bringhurst, Robert. 2018. "The Mind of the Wild." In *Learning to Die: Wisdom in the Age of Climate Crisis*, edited by Robert Bringhurst and Jan Zwicky, 7–39. Saskatchewan: University of Regina Press.

Bristow, Tom. 2015. *The Anthropocene Lyric: An Affective Geography of Poetry, Person, Place.* Houndmills, UK, and New York: Palgrave Pivot.

Buell, Lawrence. 1995. *The Environmental Imagination: Thoreau, Nature Writing, and the Formation of American Culture.* Cambridge, MA: Harvard University Press.

CBC. 2019. "Staring Down the Precipice." With shalan joudry, Cedar Meuse-Waterman, and Rose Meuse. Accessed November 30, 2021. https://newsinteractives.cbc.ca/longform/staring-down-the-precipice.

Crutzen, Paul J., and Eugene F. Stoermer. 2000. "The Anthropocene." *The International Geosphere-Biosphere Programme Newsletter* 41 (May): 17–18.

Gallant, David Joseph. 2021. "Mi'kmaq." *The Canadian Encyclopedia*, 2008. Accessed November 30, 2021. https://www.thecanadianencyclopedia.ca/en/article/micmac-mikmaq.

Hannam, Bretten, dir. 2017. *E'sekati.* Kespukwitk, L'nuekati, NS: Mazewalker Films. Written by shalan joudry.

joudry, shalan. 2014. *Generations Re-merging.* Kentville, NS: Gaspereau Press.

joudry, shalan. 2019. *Elapultiek (We Are Looking Towards).* Lawrencetown Beach, NS: Pottersfield Press.

joudry, shalan. 2020a. "How Being Still in Nature Can Remind Us of What It Means to Be Human." CBC News, May 29. Accessed November 30, 2021. https://www.cbc.ca/news/canada/nova-scotia/shalan-joudry-nature-mi-kmaw-story teller-ecologist-1.5590273.

joudry, shalan. 2020b. *Waking Ground.* Kentville, NS: Gaspereau Press.

Kimmerer, Robin Wall. 2021. "What Does the Earth Ask of Us?" Accessed November 30, 2021. https://karmatube.org/videos.php?id=8533.

Mathews, Freya. 2003. *For Love of Matter: A Contemporary Panpsychism.* Albany: SUNY Press.

Mathews, Freya. 2005. "Beyond a Materialist Environmentalism." *Nanjing Forestry University Journal* 2, no. 5: 1–5.

M'sɨtNo'kmaq, Albert Marshall, Karen F. Beazley, Jessica Hum, shalan joudry, Anastasia Papadopoulos, Sherry Pictou, Janet Rabesca, Lisa Young, and Meanie Zurba. 2021. "'Awakening the Sleeping Giant': Re-Indigenization Principles for Transforming Biodiversity Conservation in Canada and Beyond." *FACETS* 6: 839–869.

Parrott, Zach. 2020. "Indian Act." *The Canadian Encyclopedia*, 2006. Accessed November 30, 2021. https://www.thecanadianencyclopedia.ca/en/article/indian-act.

Reynolds, Ardelle. 2021. "Mi'kma'ki Storyteller Hopes Workshop Encourages 'Beautiful Increase' in Indigenous Narrative Artists." Saltwire, May 12, 2021. Accessed November 30, 2021. https://www.saltwire.com/atlantic-canada/lifestyles/mikmaki-storyteller-hopes-workshop-encourages-beautiful-increase-in-indigenous-narrative-artists-100587275/.

Snyder, Gary. 1990a. "The Etiquette of Freedom." In *The Practice of the Wild*, 3–24. San Francisco: North Point Press.
Snyder, Gary. 1990b. "Tawny Grammar." In *The Practice of the Wild*, 52–83. San Francisco: North Point Press.
Snyder, Gary. 2000. "Language Goes Two Ways." In *The Green Studies Reader: From Romanticism to Ecocriticism*, edited by Laurence Coupe, 127–131. London and New York: Routledge.
Thoreau, Henry David. 1906. *The Journal of Henry D. Thoreau*, edited by Bradford Torrey. Boston and New York: Houghton Mifflin.

5

MYTH, AFRODIASPORIC SPIRITUALITY, AND THE OCEANIC ARCHIVE IN INDEPENDENT COMICS

Paul Humphrey

Immediately prior to the Santerians' reintroduction in Marvel Comics' *The Amazing Spider-Man: Amazing Grace* series, Spider-Man steps into Don Anselmo's altar room and exclaims, "Voodoo dolls! He's a witch doctor! He's had congress with the beast!" (Molina et al. 2016).[1] The five members of this superhero team, originally introduced in the *Daredevil: Father* series, embody five individual *orishas* (spirits) from the Afro-Cuban religion Regla de Ocha, also known as Santería and Lucumí (Quesada et al. 2005). Spider-Man's outburst can be seen as a critique of the stereotypical discourse that abounds in the North American imaginary regarding Afrodiasporic religions, whereby the jocular neighborhood superhero repeatedly missteps in comprehending the structure and worldview that Santería presents for practitioners. Indeed, in response to the team's increasingly exasperated explanation that "we believe that nature binds all things together—people, animals, heaven, earth—," Spider-Man quips, "And the tree and the rock and the land and the ship—like the Force. Got it," before describing them in reductive terms as "super weather friends" (Molina et al. 2016). As this illustrates, the connections between myth, Afrodiasporic religions, and the environment have been well established in mainstream comics. Characters like Brother Voodoo—later Doctor Voodoo—and Ororo Munroe (Storm) or, on a more minor scale, Kwaku Anansi as "the first Spider-Man" (Straczynski et al. 2003) and Marie Laveau exemplify how African and Afrodiasporic mythologies have been incorporated into the Marvel Universe in ways that seek representation yet suffer variously from continued marginalization or essentialism.[2]

In recent years, Marvel and DC Comics have begun to engage more critically with these issues through series such as Marvel's *Black Panther: Avengers of the New World* (Coates et al. 2016–2018) and DC Vertigo's *House of Whispers* (Hopkinson et al. 2018–2020), to name just two. However, this essay

DOI: 10.4324/9781003348535-8

turns away from the mainstream to specific independent comics from the Puerto Rican and Haitian diasporas, as well as from Bahia, Brazil, to examine the ways in which Afrodiasporic and Indigenous spiritualities and their associated mythologies challenge hegemonic discourses about religious practice and knowledge to present decolonial futures for the region and its diasporas. Read together, the comics advance an ecocritical commentary that directly rejects colonial narratives of exploitation and extraction to then propose integrated relationships between the human and more-than-human located in the physicality of the environment. The analysis focuses primarily on the multivalent nature of the sea for Afrodiasporic peoples, a location of such deeply conflicting significance in which the many "wakes"—to cite Christina Sharpe (2016)—leave their evanescent mark. Symbolic of enslavement and mourning, spiritual and ancestral connection, and memory and multiplicity, the waters of the Atlantic and the Caribbean Sea exemplify the wider spread destruction of the marine environment. Through the narratives presented in Edgardo Miranda-Rodriguez's *La Borinqueña* (Miranda-Rodriguez et al. 2016, 2018, 2021, 2022), Greg Anderson-Elysée's *Marassa* (Anderson-Elysée et al. 2018a, 2019) and *Is'nana the Were-Spider* (Anderson-Elysée et al. 2016, 2018b), and Hugo Canuto's *Contos dos Orixás* (Tales of the Orixás; 2018), we see a turn away from the Global North to delineate Black- and Indigenous-centered epistemologies and ontologies that foreground reciprocal and mutually constitutive relationships between human and more-than-human entities.

By drawing together Brazil with Haiti, Puerto Rico, and the broader Caribbean through an analysis of Afrodiasporic spirituality and the sea, this essay is situated within a larger discussion that seeks to underscore the many "linkages between the Hispanophone, Lusophone, and Francophone Caribbean" and the wider region (Strongman 2019, 6). When incorporating Afro-Brazilian religions into his study of Afrodiasporic spirituality in the Caribbean, Nathaniel Samuel Murrell speaks directly to Brazil's "strategic ties to the Caribbean historically, culturally, and economically" (2010, 4); Roberto Strongman reiterates this point when underscoring the numerous connections between Candomblé, Santería, and Vodou as practiced in the region and its diasporas in his monograph *Queering Black Atlantic Religions* (2019). Moreover, as Kerry Bystrom and Isabel Hofmeyr observe, such a turn to the ocean and the many "routes" on and through it can illuminate "global 'inter-imperial' assemblages […] and the ways of evading and challenging empire that connect across place and time," equally as relevant in the context of the Atlantic and Caribbean Sea as it is in the Indian Ocean to which they refer (2017, 2). This troubles not only "a mode of imperial cartography that promotes the association of 'land with presence, water with absence'" (4)—as is examined below—but also the colonially imposed designations by which the circum-Caribbean continues to be framed in the academe and beyond.

Sea ontologies, memory, and the environment

In his oft-cited poem "The Sea Is History," Derek Walcott underscores the historically complicated and traumatic relationship that exists between African diasporic peoples—especially in the Caribbean—and the sea ([1979] 1986). As he demonstrates, the waters of the Atlantic and Caribbean Sea are archives of the Black experience, an assertion echoed in works by visual artists and writers such as María Magdalena Campos-Pons, Jason deCaires Taylor, Édouard Duval Carrié, and Alexis Pauline Gumbs. In particular reference to Afrodiasporic religions, scholars Omise'eke Natasha Tinsley (2018) and Toni Pressley-Sanon (2017) both explore the deep connections that exist in Haitian Vodou between the sea, history, and myth. Their work underscores the polysemy of the word "istwa," a term in Haitian Kreyòl that means both "history" or "histories" and "story" or "stories" but also, as Pressley-Sanon observes, "goes beyond them both to also encompass 'memory'" (2017, 19). Tinsley's book *Ezili's Mirrors: Imagining Black Queer Genders* focuses on the family of *lwa* (Vodou spirits) known as Ezili and on "how Vodou [...] preserves istwa (stories/histories) like that of woman-loving Danto and gender-shifting Lasirenn, stories of gender and sexual creativity that are also mythistoric records of slavery and revolution" (2018, 24).[3] Here, she draws on Keith McNeal's analysis of Afrodiasporic and Indo-Caribbean religious practice in Trinidad and Tobago, which he describes as "mythistorical archives" that

> offer imaginative truths in relation to which people make sense of and compete over life; apprehend the present, past, and future; [...] legitimate or mystify social relations and inequities of power; and seek self-transformation or greater forms of collective change.
>
> *(2010, 188)*

This interweaving of myth and history that occurs in Haitian Vodou and the spiritual traditions McNeal examines equally applies in the context of Cuban and Puerto Rican Santería and Brazilian Candomblé.

"Water overflows with memory," M. Jacqui Alexander writes in her exploration of "Pedagogies of the Sacred" (2005, 290). Such memories of celebration, trauma, and resistance are written onto the sea through the embodied spiritual practices of Afrodiasporic and Indigenous communities, through rites and rituals that implicate the body and the waters as conduits of spiritual and ancestral connection. As Alexander recounts, many reformulations have taken place in "the Crossing"; for example, Yemayá—*orisha* of the sea and maternity—and her associated *orixá* Yemanjá "[reign] in Candomblé and Lucumí, assuming the position that had been accorded her River sister Oshún in Yorubaland," while Lasirenn, Agwe, Ezili Dantò, and Ezili Freda fulfill this role in Haitian Vodou (2005, 292).[4] Framed in another fashion, then, the sea serves as McNeal's "mythistorical archive." The histories, spiritualities, and lived experiences of enslaved individuals—both those "who did not survive the holding and the sea" and those who did—are left as traces on and in the water, "with us still, in the time of the wake" (Sharpe 2016, 19). Through "wake work," it is possible to "imagine ways of knowing that past, in excess of the fictions of the archive," and simultaneously "recognize the many manifestations of that fiction and that excess, that past not yet past, in the present" (13). The sea thus becomes that site of engagement and inquiry, resistance and remembrance, a fluid archive of "bodily memory" and "sacred memory" that challenges colonial and patriarchal notions of time, history, ancestral connection, and ways of knowing and being (Alexander 2005, 290).

Engaging the sea as epistemology and ontology is critical for understanding how cultural production from the Caribbean and its diasporas perceives current environmental crises the region faces. From framing the sea as a site of relation in works by Édouard Glissant, Antonio Benítez Rojo, and Paul Gilroy, to Kamau Brathwaite's (1983) and later Elizabeth DeLoughrey's (2007) concept of "tidal dialectics" or "tidalectics," scholars of the Caribbean have sought to theorize the complexities inherent in the cultural, political, and historical symbolism ascribed to the waters that divide and connect the islands and littoral areas of the region. More recently, DeLoughrey has developed the concept of "sea ontologies," derived from the "oceanic imaginary" in current cultural production "concerned with our watery futures" (2017, 34–35). As she writes, the Atlantic is "an oceanic archive that lacks the place-based narrative and rituals of memorialization" and thus, in this context, sea ontologies "might characterize the connection between ancestry, history, and non-Western knowledge systems in submarine aesthetics" that we see in the Caribbean, Brazil, and beyond (35). No longer "blank space or *aqua nullius*," these waters are imagined and depicted as "a multispecies and embodied place in which the oceanic contours of the planet, including its submarine creatures, are no longer outside the history of the human" (42). In elaborating the concept of sea ontologies, DeLoughrey turns to abovementioned artist Jason deCaires Taylor's underwater museums and sculpture parks, the Caribbean iterations of which are installed in shallow waters around Grenada; Isla Mujeres near Cancún, Mexico; and New Providence and Musha Cay, The Bahamas. Taylor's submerged sculptures are "transformed by salt, currents, pressure, and the rapid occupation by multispecies ecologies" (37), ultimately becoming sites that spotlight and anthropomorphize the crisis faced by tropical coral reefs which, for DeLoughrey, is "one of the most visible indicators for the Anthropocene" (40). Such anthropogenic destruction is caused by fluctuations in ocean temperatures (which also contribute to hurricane activity in the region), acidification, runoff from agricultural land, fishing techniques, and tourism, among other human activities.

In this way, the oceans continue to be subject to "hydrocolonialism," coined by Hofmeyr (2022) to encompass "various forms of maritime imperialism […], colonization of water […], the ship as a miniature colony […], flooding of occupied land […], and colonization of the idea of water" (15–16). Indicative of the "oceanic turn" in scholarship of the environment in recent decades, concepts such as *aqua nullius*, hydrocolonialism, the oceanic archive, and sea ontologies seek to critique and dispute the "imperial cartography" that portrays oceans as "bounded physically or historically," to acknowledge the waters' "human-nonhuman-material-discursive connections that challenge our notions of time, space, culture, history, and humanity" (Shelton 2019, 152). It is to these multifaceted connections and their complex spiritual, epistemological, and ontological meanings that this essay turns, focusing on the many *istwa* written in the wake that—when further elaborated—offer the potential for decolonial futures based on reciprocal relationships with the physical and spiritual environment. As Fernández Olmos and Paravisini-Gebert observe, a close, mutually sustainable relationship with the environment is fundamental to Creole religions of the region such as Santería, Candomblé, and Vodou, and in recent years "these spiritual connections […] have become the foundation for ethical and moral choices about community and the environment" (2022, 36).

Within this framework, recent decades have seen numerous works of speculative fiction and comics published that explore the relationships between these and associated spiritualities and the sea. In independent comics, current publications from both sides of the Atlantic include Cameroonian and French creator Reine Dibussi's *Mulatako* (2017, 2020),[5] contributions to the anthologies *Mañana: Latinx Comics from the 25th Century* (Gil 2021) and *Gwan: Volume 2* (Walford 2020), and Jiba Molei Anderson's (2002–) depiction of the *orishas* from Yoruba religion in the ongoing *The Horsemen* series. Although very different in genre, Canuto's *Contos dos Orixás* explores a closely connected spiritual mythology, that of Brazilian Candomblé. Retelling the *ìtàn* or myths of the *orixás*, the graphic novel is written in Brazilian Portuguese with Yoruba terminology throughout and depicts the *orixás* in a timeless past when "divine beings walked with men, passing between Orum [the realm of the *orixás*] and Aiyê [the physical world]" (Canuto 2018).[6] Through the *orixás* Yemanjá and Oxum, the reader witnesses a reverence and deep concern for the waters that are their realms and serve as a key source of restorative energy throughout the text.

Axé, archives, and oceanic futures

Prior to publishing *Contos dos Orixás*, Hugo Canuto produced a series of 14 comic book covers depicting a single *orixá* per cover, using their characteristics, associated colors and objects, and powers drawn directly from their spiritual counterparts to create his suite of Candomblé superheroes (see Canuto [2017]). Accompanying the series of portraits was an adaptation of Jack Kirby's cover for Marvel's *The Avengers* #4 (Lee et al. 1964); Canuto's version replaces the returning Captain America and the other Avengers with his new team, "The Orixás." In addition to the witty homage to Kirby's influence on his own style, Canuto substitutes Marvel's Caucasian characters with equally influential figures from Afro-Brazilian and West African mythology. Importantly, the source of the characters' powers in *Contos dos Orixás* is *axé*, "the vital life-giving force of the material and spiritual worlds" through which "equilibrium with the world of the spirits [...] is maintained" (Murrell 2010, 180).[7] As the comic narrates at its opening, when the creator Olorum "parted the gourd of existence[, f]rom it sprouted axé, [...] whose energy spread through the void, creating stellar oceans and forges of incandescent galaxies" (Canuto 2018). Oceanic imagery abounds from the outset, with the opening full-page panel in blue and white referencing these "stellar oceans" of Orum.[8] Overleaf, while Obatalá molded the land and created humankind, Yemanjá "remained outside the movements of the world above, entertained by the music of the corals, [...] populating the sea with colorful shoals" (Canuto 2018; see Figure 5.1).[9] Hearing the lament of her enslaved children, Yemanjá's tears caused a storm, washing over the land and creating a valley with a "river of golden and fertile waters," where Cidade Mãe (the Mother City) was later built (Canuto 2018).

In this three-page sequence of panels, the gutters are deep blue—the color of the stellar and earthly oceans—and the colors used are predominantly hues of blue and yellow. Frederick Luis Aldama has described the gutter as the "*sine qua non* shaping device of comics, [...] the space for us to imagine movement, thought, and feeling" (2017, 2), observations that fit with other scholars' theorizing of the gutter as a space that fractures time and as a site of memory (see, for example, McCloud [1993, 66–67] and Chute [2016, 35–36]).

FIGURE 5.1 Panel from *Contos dos Orixás*, Hugo Canuto. © 2018 Hugo Canuto www.hugocanuto.com and @hugocanuto_art

Here, then, the narrative of creation, oppression, destruction, and regeneration the panels recount is both contained in and transmitted through the blue of the gutters; be it the metaphorical ocean of the skies or the watery realm of Yemanjá, the sea is that space of memory and movement, thought and feeling. Added to the blues of Yemanjá and the white of Obatalá that denote their presence throughout, Canuto's use of yellow here also invokes Oxum, *orixá* of "sweet, fresh water and sensual love" (Murrell 2010, 173). Appearing as a character much later in the narrative, Oxum presides over Cidade Mãe and the *axé* of Yemanjá's waters that it contains. However, incorporating her into the aesthetics of the comic at the outset underscores the importance of the rivers and seas—and the *orixás* associated with them—as the means through which these stories, histories, and mythologies are told. It also reiterates the close relationship between the two *orixás* and the fundamental roles they play in Yoruba spirituality and the Candomblé of Bahia. As Canuto (2021) has explained, the visual rhetoric associated with both belief systems is integrated into the comic; examples include a statue of Yemanjá in Salvador, Bahia, used to inspire the *orixá*'s visual form, the imagery of individuals crossing the sea found on ornamental wooden doors from Yorubaland, the incorporation of designs from traditional Yoruba cloth, and using real-life models from Salvador for his characters. In so doing, he builds an aesthetic that captures the complexity of the transoceanic connections that exist between these two spiritual systems (Canuto 2018, 2021). Indeed, the same symbolism and color choices from these first pages reappear when Exú, trickster *orixá* who opens the portal between Orum and Aiyê, leads the assembled team of warriors along the yellow path that traverses the crossroads, a timeless, blue "sea of night" where gourds harbor "axé in its purest form" (Canuto 2018).

The transition from the timeless past of the opening to the narrative present—itself still temporally set in a time previous to our own—sees the style change to a more contemporary aesthetic with standard white gutters. The plot details the rise of the Ajogun, the "adversary forces of humanity," and their march to take Cidade Mãe under the leadership of the fictional Ajantala, while the *orixás* and other members of the team seek to thwart the attack and protect the realm of Yemanjá and Oxum (Canuto 2018). Inspired by the shrine to the Yoruba *òrìṣà* Ọ̀ṣun in Oṣogbo, Yorubaland, the "Mother City" is fashioned out of gold and situated in the middle of a river, a direct reference to Oxum/Ọ̀ṣun in both material and location (Canuto 2021).[10] In addition to the yellow associated with Oxum from above, the *orixá* arrives clad in gold and blue, riding the waters in a golden vessel and carrying the ritual fan held by Yemanjá in Figure 5.1, all set against a sky tinged with pink and so combining in these panels the three colors Canuto (2021) has associated with this *orixá*.[11] As above, the water here is a site of power and knowledge, the source of the *axé* that is found throughout the city. While touring Cidade Mãe, Oyá comments that the Ajogun's intention must be to steal its gold, underscoring that the parallel history of subjugation and enslavement the comic depicts also serves to reject the European colonial project. Notably, Oxum counters that—rather than the gold—the *axé* found "among sacred leaves, in the heart of a star, or in the grains of sand," the "richness that we protect," is much more valuable (Canuto 2018). "If the source that resides in the waters falls into the wrong hands," she continues, "it puts the equilibrium of life in Aiyê at risk."

Earlier in the comic, Ajantala claims that "The gods will be forgotten," making this assertion as an entity external to the Candomblé worldview (Canuto 2018). For Oxum, however, the threat is not only material (the gold, the water) and epistemological (the histories and myths written into the waters); it is also existential, endangering life throughout the earthly plane. Glimpsing Ajantala's niece in the mirrored waters of Cidade Mãe, Oxum reaches through the water to bring her to the Mother City where, together with Oyá, they disperse the mist Ajantala uses to control the Ajogun and thus halt the invasion, leaving Xangô to dispatch the aggressor. It is by gathering the knowledge and power imbued in the waters and calling on the "ancestral mothers" that the Ajogun are stopped and equilibrium in Aiyê is restored (Canuto 2018). Focusing on *axé, Contos dos Orixás* offers a perspective deeply grounded in the myths and histories of Candomblé that rejects a (hydro)colonial, extractivist relationship with the environment—founded on the notion of *aqua nullius*—in favor of reciprocity, recognizing the power and agency of the waters. If Canuto's retelling of the *ìtàn* is situated in a timeless past with contemporary resonances for our relationship with the marine environment, then Greg Anderson-Elysée's *Marassa* (Anderson-Elysée et al. 2018a, 2019) draws on the *istwa*—to borrow the Kreyòl term—of these waters to project potential Afrocentric and decolonial futures.

Although only two issues of *Marassa* were published before the title paused production, this "Afrofuturist space opera" that brings together "space pirates, bounty hunters, humanoid snake doctors, cosmic Vodou, and breathing wooden mask babies" resonates with Canuto's narrative (Anderson-Elysée et al. 2018a). The comic takes its title from the twin *lwa* of Haitian Vodou, the marasa, from which the fraternal twin protagonists' names Mara and Sa are derived. In the narrative present, Mara is a princess on an unnamed planet whose queen and crown prince have the infinitesimal cosmologies of space depicted on their skin, while Sa continues the twins' original joint enterprise as space pirates, joined by his adopted son and "wooden mask baby," Petit. As Anderson-Elysée has stated, the comic demonstrates the many transregional connections the creator perceives in the contemporary moment (BlackSci-Fi.com 2019). On the one hand, this is exemplified in the comic's use of French, Spanish, and Haitian Kreyòl in its mostly English-language narrative, achieved through character names (e.g., Koulèv [Kreyòl] and Luna [Spanish]), locations (e.g., Planet Vè'vètè [Kreyòl] and Sankofa, a concept from Akan mythology), and a small number of minor dialogue exchanges. Most notable for this analysis, however, is that the diegesis and artwork draw heavily on interconnected Afrodiasporic spiritualities, not least in the conceptualization and visual presentation of characters. For the purposes of this discussion, the most salient characteristics include the comic's use of colors—similar to Canuto above—and the incorporation of the ritual designs for the *lwa* into its visual rhetoric; the concept in Vodou that the marasa comprise not only two but also three entities; and the importance of transcorporeality as a framework for understanding possession performance in these Afrodiasporic religions.

The marasa function as a key organizing principle of the comic. In Vodou, the spiritual power of the twins is derived from the strength of the bond that exists between them, which in *Marassa* is reiterated on multiple levels. Beginning with the cover, the source of power visually emanates from Mara and Sa's joined hands, and the partial *vèvè*—elaborate cornmeal drawings that both represent and propitiate the *lwa* during Vodou ceremonies—for the marasa is incorporated into the title's lettering. These strong ties are repeated throughout: the same energy radiates from Sa and Mara's clasped hands as he entreats her to join his quest for their "heirloom," Kavo's Bones (Anderson-Elysée et al. 2018a); the golden yellow of Mara's outfit in issue one is replicated in Sa's bandanna and the designs on his jacket; the purple of Sa's jacket throughout the series reappears in Mara's clothing at the close of issue two. Additionally, as Pressley-Sanon notes, the marasa twins have a close association with Legba, the guardian of the crossroads and counterpart in Vodou to Exú (2017, 15). After Sa's death at the end of issue one, his lover Shelly searches for his twin spirit in the waters surrounding her bar, Sankofa; as she descends, Legba's *vèvè* appears in her eyes, prefiguring her encounter with the *lwa* himself. In *Contos dos Orixás*, Exú constructs the crossroads in order to traverse the "stellar oceans" between Orum and Aiyê, thereby facilitating passage for the *orixás* to the earthly realm. Here, however, the depths of the watery seas constitute this threshold, in which the realms of the physical and spiritual, the living and the dead intersect.

A descent *anba dlo* (underwater) is multiply significant in Haitian Vodou, with the depths of the ocean constituting the passage to "the mythical Ginen," where the *lwa*, Gede spirits, and the dead reside (Pressley-Sanon 2017, 100).[12] Additionally, for those devotees whose initiation into Vodou involves a period spent *anba dlo*, the water transports you "beyond words, beyond space-time, beyond self" as you drift, before reaching the clarity of knowledge that initiation provides (Tinsley 2018, 146).[13] Shelly seeks this *konesans* (spiritual and ritual knowledge) and, in a series of wordless panels that replicates this suspension of time and sense of self, she engages different spiritual traditions of the region, divining Sa's death through a card reading and collecting symbolic gifts to propitiate the *lwa* and *orisha* in exchange for their assistance. First she encounters Legba, in a red and black tunic, followed by his Santería counterpart Eleggua, who wears the black and red beaded bracelets and necklace that represent this *orisha*. Together, Legba and Elegguá accompany Shelly across the underwater threshold to meet representative entities of several Vodou and Santería spirits. The pink sky of Oxum in *Contos dos Orixás* becomes the pink waters here, a stark contrast to the black of the interplanetary vacuum in which Sa and Petit were captured. Bathed in the presence of Ezili Freda—Oxum's counterpart in Vodou, also associated with pink—Shelly encounters a mermaid who embodies these associated spirits: her yellow tail evokes Ochún and Oxum, the design where her torso and tail meet is reminiscent of a stylized outline of Ezili Freda's heart-shaped *vèvè*, and the gold jewelry she wears is associated with all three.[14] Registering Shelly's pain for her lover, this spirit of love, sexuality, and sensuality then transports her by means of an intimate kiss to Ginen and the presence of the Gede spirits Bawon Samdi and Gran Brijit (see Figure 5.2).

This intimacy that serves as the portal for Shelly between the watery realm of the spirits and Ginen can be read as analogous to the process of trance possession in Vodou, Santería, and other associated spiritualities. As Strongman (2019) outlines, possession in Vodou sees one constituent element of the devotee's self—the *gwobonanj*—displaced by the *lwa*, the latter of whom performs their subjectivity on the individual's physical body in a process Strongman theorizes as transcorporeality.[15] This process is exemplified in *Marassa* through the kiss, where Shelly and the spirit become an identifiable third entity in Ginen. As Figure 5.2 shows, behind the table where Bawon Samdi plays checkers stands a child-like character who, in the next panel, wears the mask of Ochún/Oxum/Ezili, Shelly's necklace, and both characters' earrings. Sartorial choices again indicate familial and spiritual connections: Shelly's black bathing suit is a color associated with the Gede, which in conjunction with a red bandanna evokes Legba and Elegguá, and the gold jewelry references Ochún/Oxum/Ezili. Most important, Bawon Samdi's opponent wears an ankle adornment that reveals his own identity. Decorated with the curlicues seen on both Sa's and Legba's jackets, which also appear on Sa's skin when he blasts green-colored energy from his hand in issue one, it appears that Shelly has located her lover's spirit, combined with Legba, who had opened to her the crossroads between the physical and spiritual realms. Here, in the mythical space of Ginen, both children are transcorporeal entities, through whose bodies perform the associated—or might we say twinned—subjectivities of Sa and Legba, Shelly and Ochún/Oxum/Ezili.

This idea of the third entity is intrinsic to the concept of twins in Haitian Vodou. The marasa has a third iteration, called the marasa twa, and the third "twin" born after the other two, the "dosou" or "dosa," increases the twins' already potent strength (Pressley-Sanon 2017, 15). This third entity, Alessandra Benedicty-Kokken observes, "serves as an arbitrating force in the worlds of persons and communities constantly downtrodden by globalizing forces from slavery to the neocolonially inflicted environmental forces of destruction" (2015, 232). In *Marassa*, the reader sees multiple groupings of three, be

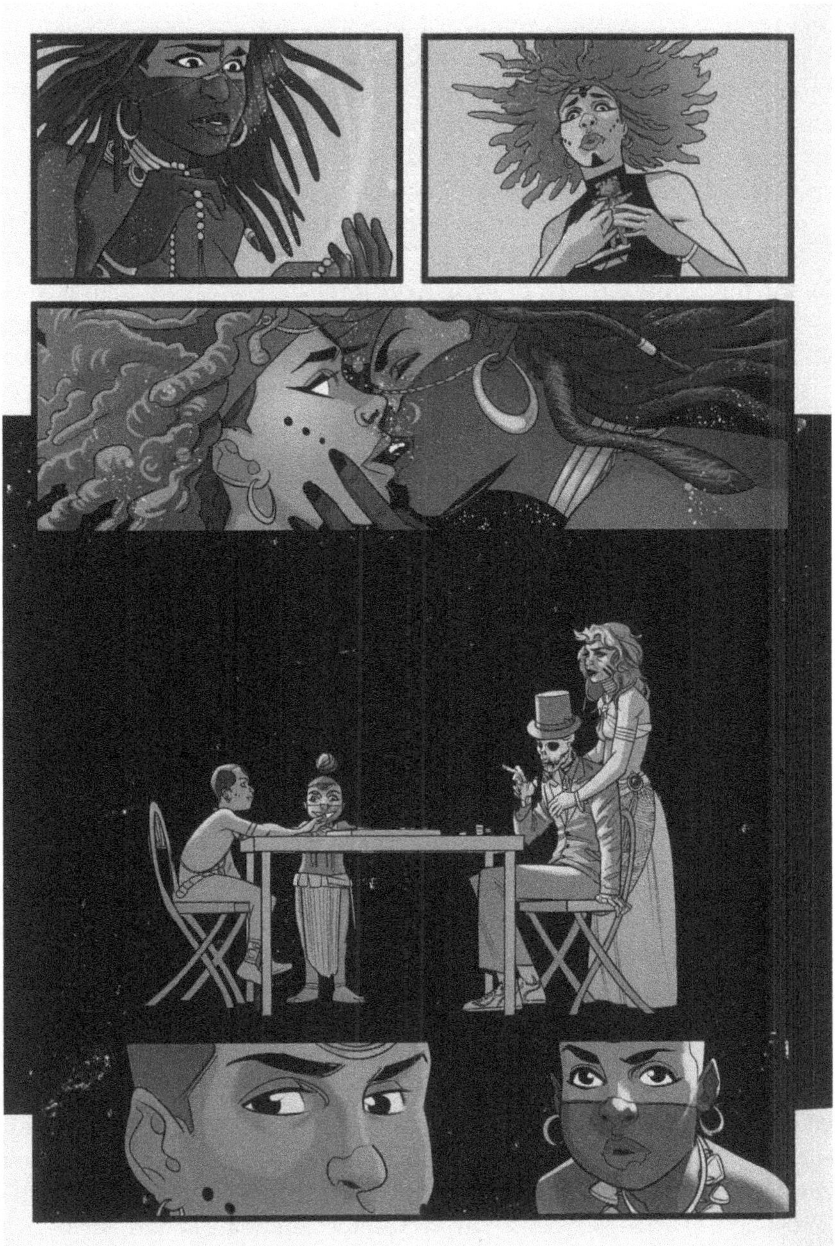

FIGURE 5.2 Panels from *Marassa* #2, Greg Anderson-Elysée. Art by Antonello Cosentino and Francesco Montalbano. © 2018 Greg Anderson-Elysée; published by Evoluzione Publishing www.evoluzionepublishing.gumroad.com

it the relationship between Mara, her husband, and her mother-in-law; the composition of the transcorporeal children in Figure 5.2; or the organic connections between Sa, Petit, and their wooden spaceship, Bwa (itself a commentary on the reciprocal relationship between Vodou and the natural environment).[16] However, the most salient iteration of the marasa twa is Mara, Sa, and Shelly, whereby Shelly becomes the dosa, an embodied example of "one plus one equals three," to use the oft-cited description of the marasa twa (Bellegarde-Smith 2005, 11). The artwork gestures toward this role: the three dots on each of her cheeks through both issues contrasts with the two on the cheek of Sa/Legba in Ginen (Figure 5.2). Shelly also appears with two further black markings on her face throughout this sequence, one on her forehead and the other on her chin; between these is situated the two sets of three dots, demonstrating in multiple forms the position she holds. Pressley-Sanon describes the dosou/dosa as the "attendant third element/space/moment of resolution or completeness through creativity," a role Shelly fulfills in locating Sa to reestablish the connection between the fraternal marasa (2013, 119). In resolution, at the issue's close, Mara is dressed in the purple associated with Sa, propped up on a seat whose purple cushions form the bones of the dead often seen in depictions of the Gede, while Shelly—with a further two sets of three markings on her breastplate—pilots the ship to rescue the missing members of their multiple marasa twa.

Framed thus, the sea constitutes this third space, the site of "resolution or completeness through creativity" (Pressley-Sanon 2013, 119). In *Marassa*, the water takes the reader "beyond words," to quote Tinsley from above, in a five-page sequence of panels without captions or dialogue, where the colors and details of the artwork convey the *konesans* Shelly seeks. This polysemous site of trauma and possibility is the space in which the (hydro)colonial notions of extraction and expendability of characters of African descent are challenged, through the transcorporeal process of trance possession. In discussing the philosophical ramifications of transcorporeality, Strongman writes that the labeling of colonized and enslaved Black bodies as "soulless [so] as to close off their philosophical corporeal openness" went hand-in-hand with "legislatively prohibiting those religious rituals of trance possession that render black bodies inhabited or soulful" (2019, 4). Beyond this spiritual context, Stacy Alaimo theorizes transcorporeality "in an oceanic sense," describing a "mode of being" that is not "a finished, self-contained product of evolutionary genealogies but a site where the knowledges and practices of embodiment are undertaken as part of the world's becoming" (2016, 127). Conceived thus, "bodies, things, and objects" are not "separate entities, but instead [...] already part of intra-active networks and systems that are simultaneously material, discursive, economic, ecological, and biopolitical" (133).[17] As transcorporeal assemblages in both Strongman's and Alaimo's formulations, the characters and associated spirits in *Marassa* do the "wake work" Sharpe charges, recognizing the *istwa* of death and spiritual and ancestral connection found in the waters surrounding Sankofa, itself an Akan symbol representing the intertwined nature of past, present, and future. The timeless "stellar oceans and incandescent galaxies" of *Contos dos Orixás* become, here, a future sea crossed by space pirates and bounty hunters, tracing the histories/stories of oppression and possibility, division and connection on its celestial waters. As narratives steeped in Afrodiasporic spiritualities, not only do these comics depict in part the oceanic archive DeLoughrey invokes, they also detail nuanced relationships between the human and more-than-human that might serve as a model for decolonial engagements with the environment.

Lòtbò dlo: watery webs of ancestral connection

At the beginning of Anderson-Elysée's *Is'nana the Were-Spider: Vol. 1*, the titular character's father counsels, "The tapestry of knowledge, of stories from the blessings of Nyame himself, we spin the webs together, my son" (Anderson-Elysée et al. 2016). Using the stories of the trickster spider god Anansi from Akan mythology as its inspiration, the comic depicts the adventures of Anansi's son—whose name is his own in reverse—once he leaves the spiritual realm known as the Mother Kingdom for the physical plane of human existence, thereby retelling these myths while writing new narratives for the 21st century. Though not located in the sea, this comic looks to the many spaces *lòtbò dlo* (across the water), weaving literal and metaphorical webs between African and Afrodiasporic spiritualities and folklore in the abovementioned multivalent space of the Crossing. For John Jennings, "Anderson-Elysée's wonderful remixing of African tales from the diaspora [...] represents a powerful retelling of our history [...] through a much darker lens of horror" ([2019]). Despite the different genre of this comic, the power imbued in Anansi and Is'nana is similar in source to the multiple marasa twa of Anderson-Elysée's *Marassa*, since it derives in large part from the manifold connections between father, son, and their fellow spiders. Moreover, the comic forces readers to grapple with issues of coloniality, extractivism, and the environment through a character who—as a were-spider "son of a god" (Anderson-Elysée et al. 2016)—transmogrifies between human, spider, and a compendium of the two, collapsing divisions between human, nonhuman, and more-than-human.

Within this frame of reference, *Is'nana* is similar to *Contos dos Orixás* and *Marassa* in the plurality of linguistic influences it incorporates. First, the names of figures from Akan mythology remain unchanged, and additional characters in the Mother Kingdom are given names in English (e.g., Witch Mistress Five), Haitian Kreyòl (e.g., Zarenyen [spider]), and Yoruba (e.g., Oloko), signaling the connections between different mythologies within the spiritual realm. Notably, the white gutters and distinct art style that designate the Mother Kingdom in volume two indicate that its first chapter continues the identically named "In the Beginning" chapter that closed the previous book.[18] In a series of panels that sees the hornets, wasps, and bees unite to rescue the captured Mmboro, Anderson-Elysée's strategy for naming reflects these ties between West Africa and the Caribbean. The emissaries from the hornet and bee colonies—presided over respectively by Queens Mmboro and Uh-yen (similar to the Yoruba word for honeybee or honey, "oyin")—approach Queen Gep (Kreyòl: "wasp") to create an alliance. Successful in their petition, Kantite (Kreyòl: "quantity" or "numerous") and Myel (Kreyòl: "honeybee") rescue Mmboro, prior to Kantite's transformation in the physical plane into the many-bodied hornet daughter of Mmboro that her name suggests.[19] This politics of naming—that is, using terms from both African and Caribbean languages for the matriarchs—highlights the transatlantic ancestral connection while underscoring that Afrodiasporic mythologies are not solely derivatives of their West African counterparts.

The issue of a fundamental fracturing of the relationships between humans, nonhumans, and the more-than-human that this chapter reprises is raised by Is'nana in "Forgotten Stories," at the beginning of volume one. In the urban environment, the protagonist laments the absence of "the tall trees and the colors illuminated by the hot sun" and the lack of a visceral connection with the land, concluding that, "In this place, I see humans walking and spending their days being chipped further away from their essence…" (Anderson-Elysée et al. 2016). The consequences of humanity's deteriorating relationship with the environment are laid bare by the queens' emissaries, who counter Gep's indifference to their dwindling numbers with "Queen! -bzzt- Once the beez depart from nature, we, the hornets and wasps, are next" (Anderson-Elysée et al. 2018b). Although the hornets retaliate in the physical plane by taking over human bodies to convert them into the "shells" they inhabit, thereby becoming the main antagonists, the comic asks its readers to consider the question of reciprocity that Is'nana outlines, that the vengeful actions of the hornets are a reaction to the "fear our kind will soon be extinct." The protagonist's solution is to further develop the "webway" that saw him first transported to the physical realm, constructing an interdimensional network of African and Afrodiasporic mythologies focused on the physical and mythical presence of spiders and trickster spirits on both sides of the Atlantic. At first, webs are a means for Is'nana to borrow his fellow spiders' eyes, with each sector a window to his siblings' current viewpoint. However, when passing through the portal from the Mother Kingdom, he glimpses numerous mythological trickster spirits: Legba at the crossroads, marked by his *vèvè*; Papa Bois and La Diablesse, spirits of the trees from Caribbean folklore; the Kishi figure from Bantu mythology; and Br'er Rabbit of the southern United States (Anderson-Elysée et al. 2016). Anderson-Elysée also incorporates his own trickster character, Jeremy Tableau, who inhabits a dream world presided over by the Greek god Morpheus. Whereas in volume one Is'nana is unable to control his passage through the portal, in volume two and beyond he can enter the webway at will, navigating between physical spaces, interweaving mythologies, and using spiders' memories to counter the numerous threats faced.

Subsequent to these volumes, Anderson-Elysée has published several one-shot issues and contributions to anthologies featuring Is'nana. These continue to reimagine intersections between figures from African folklore and spiritualities in conjunction with Afrodiasporic and Indigenous folktales and mythologies of the Americas. In a prequel issue, for example, he "writes of Is'nana, Lil Br'er, Ndua, Coutee, and Tumfeh together as friends facing challenges as a team, [...] weaving a golden thread of cultural continuity and connection through the global black experience" (Hollman [2022]). Elsewhere, figures Is'nana encounters include the Central and West African gods Xevioso and Oshún (Anderson-Elysée et al. [2021a]), the South African Impundulu (Anderson-Elysée et al. [2021a]),[20] gods from the Igbo, Maasai, and Yoruba peoples (Anderson-Elysée and Jones 2022), and Bohpoli and Impa Shilup from Choctaw mythology, alongside whom appears the Vodou spirit Gede Nibo (Anderson-Elysée et al. [2021b]). Returning to volume two, a brief interlude in "The Hornet's Web" chapter becomes a new narrative arc, in which the trickster spirits Anansi and Legba meet at the crossroads—marked again by the latter's *vèvè*—and attempt to locate Is'nana and Legba's dog Cerberus (Anderson-Elysée et al. 2018b).[21] Anansi and Legba find their charges in the world of Haitian folklore with Uncle Bouki and the trickster Ti Malice, where Jeremy Tableau also emerges. Conflating the mythological with the comics genre, Cerberus leads Jeremy and Is'nana through a series of portals to the worlds of six other independent comics and so expands the reach of the earlier webway and continues the creation of intersecting narratives for the 21st century. This is achieved through a character who, by his very nature, troubles the colonial epistemological division in the Caribbean between humans and the environment, weaving these interconnected webs of mythologies and spiritualities. In their introduction to *Postcolonial Ecologies: Literatures of the Environment*, DeLoughrey and Handley write, "The ecocritical interrogation of anthropocentrism offers the persistent reminder that human political and social inequities cannot be successfully and sustainably resolved without some engagement with the more-than-human world" (2011, 25). A necessary part of these postcolonial ecologies they delineate is a nuanced engagement on both local and global scales, one that "avoids facile generalizations" while recognizing the value in "imagining relations in new ways in order to forge new epistemologies" (29). In tandem with *Contos dos Orixás* and *Marassa*, then, *Is'nana* exemplifies this local yet global sensibility, asking the reader to engage with its writing of mythologies past, present, and future, albeit through what might seem the unlikely genre of horror.

The final title examined here, *La Borinqueña* by Edgardo Miranda-Rodriguez (Miranda-Rodriguez et al. 2016, 2018, 2021, 2022), has received the most critical attention. The series currently comprises three issues and two one-shots, alongside the protagonist's numerous appearances in *Ricanstruction: Reminiscing and Rebuilding Puerto Rico* (Miranda-Rodriguez 2018) and a sequence of four minicomics on Chocolate Cortés wrappers. Following its eponymous Afro-Nuyorican superhero whose powers are derived from her connection to the Taíno spirits of motherhood and of the sea, wind, and mountains, *La Borinqueña* sees its protagonist fight alongside activists resisting colonial and neocolonial powers in Puerto Rico, not least in reference to issues of environmental concern. These include the consequences of hurricanes and earthquakes for the archipelago, the impact of tourist development of the coastline, the unregulated dumping of hazardous waste, and the deleterious effects of human activity for turtle nesting and on bird, frog, and bat populations. In addition to the many journalistic articles and interviews available in online and print media, academic analyses of the comic include the representation of the Afro-Puerto Rican diaspora (I.M. García 2018), narratives of indigeneity and Taíno spirituality (E. García 2020), and my own examination of how issues one and two depict decolonial futures, spirituality, environmental crises, and resistance in the aftermath of Hurricane María (Humphrey 2021).

As Enrique García demonstrates, "La Borinqueña is empowered by the island [...] to protect its people, but this power is built on the foundation of Taíno culture, and does not stem from the two oppressive Eurocentric traditions" (2020, 226). First, this is exemplified by the superhero's name, which references both the original Taíno name for the island, Borikén, and the 1868 poem "La Borinqueña" by *independentista* poet Lola Rodríguez de Tió that, later set to music, became Puerto Rico's original national anthem. The comic represents "Puerto Rican culture as racially and ethnically mixed but positions the Taíno heritage at the center of the model," drawing on the Afro-Boricua superhero's connection to the Taíno spirits based on a mutual "political ideology of resistance as well as [...] a shared racial, cultural, and geographical experience" (E. García 2020, 229–230). Through associations with the *orishas* Yemayá and Changó, along with the importance of the ceiba tree in both Santería and Taíno spirituality, La Borinqueña serves as a conduit for Afrodiasporic spiritualities, too (see Humphrey 2021, 67–69). Most notable for this analysis is that not only is the superhero connected to Yucahú—the Taíno "spirit of the sea and mountains"—and Huracán—"spirit of the storms"—but also, as the name of her alter ego Marisol Ríos de la Luz ("Sea and sun [...] rivers of light") indicates, she embodies this connection between the waters of the sea and the oceans of the sky examined above (Miranda-Rodriguez et al. 2016). Indeed, Marisol's powers are revealed when she connects the five crystal tears of Atabex found while conducting research in the caves of Puerto Rico to form "la Estrella del Camino" (the star of the pathway). Transported to a timeless dimension, visually presented against a dark, star-filled sky, La Borinqueña is led to the Taíno spirit of motherhood by means of the waters, in a physical and spiritual journey that mixes fresh river water with the metaphorical salt water of tears. In addition to Atabex being "the ancient spirit of your deep past [...] the water that flows through your consciousness," this confluence suggests the presence of Yemayá and her sister spirits that is explicitly revealed in issue two (Miranda-Rodriguez et al. 2016). After being enclosed in the star at the end of this issue, La Borinqueña must navigate the multiple histories, stories, and memories of resistance—past, present, and future—and develop the *conocimiento* (akin to *konesans*) required to prevent the ongoing extraction of power from la Estrella del Camino and all it represents.

At the opening of issue three, La Borinqueña encounters Opiyelguobiran, "the Taíno cemí that guides the souls of lost lives to Coaybay, the realm of the spirits" (Miranda-Rodriguez et al. 2021). As Figure 5.3 shows, Opiyelguobiran is also represented in the form of a dog (a comparison found in 15th-century Spanish missionary Ramón Pané's writings [1999, 28]) and accompanies the spirits of La Borinqueña's Taíno and African-descended ancestors from Coaybay to what the reader now learns is Koeia, the timeless dimension inside la Estrella del Camino which takes its name from the Taíno word for star (9).[22] La Borinqueña seeks her escape and Cacique Mabodacama, with whose spirit the protagonist interacted in issue two, guides her through a series of ancestral encounters to develop the *conocimiento* she needs. Such knowledge is gained through both spiritual connection and ritual practice, one step of which is La Borinqueña dancing the spiritually significant Bomba, accompanied by the *cimarrones* (maroons) whose spirits she had previously witnessed at the ceiba tree. After encountering the vejigante figure of Puerto Rican folklore followed by notable individuals from 19th-century resistance movements, La Borinqueña glimpses in the star's crystal facets the events and experimentation befalling her friends and fellow activists in the physical plane. She sheds a single tear, the salt water once again connecting her with the spirit(s) of motherhood, and the *conocimiento* she has developed enables her to return to Borikén. Over several sequences of panels

FIGURE 5.3 Panel from *La Borinqueña* #3, Edgardo Miranda-Rodriguez. © 2021 Somos Arte www.la-borinquena.com

that narrate Puerto Rico's history of resistance, Miranda-Rodriguez therefore locates this "mythistorical archive" in an ancestral dimension that connects the physical and metaphorical waters of Indigenous and Afrodiasporic spiritualities and folklore.

As in the other comics discussed, Miranda-Rodriguez seeks the power of myth, historiography, and spirituality as a means to construct decolonial narratives that combat the effects of colonial environmental destruction in Puerto Rico. He writes trauma, resistance, memory, and possibility into a multivalent sea space—both celestial and oceanic—that interweaves mythologies and recognizes the necessity of reciprocal and therefore sustainable relationships with the environment if Puerto Rico and the wider region are to mitigate the current crisis. The superpowered individuals introduced in this issue derive their powers from the flow of light connecting them to La Borinqueña and la Estrella del Camino, and four of these characters team up with La Borinqueña to form the NiTainos (Miranda-Rodriguez et al. 2021, 2022). Iguaca's and Oro's superpowers reflect characteristics of the Puerto Rican amazon and the *coquí* tree frogs, both of which have suffered population decline due to anthropogenic factors. Meanwhile, the issue's newly created human–bat antagonist, La Gárgola, forces readers to consider the consequences of increased urbanization and hurricanes on Puerto Rico's bat population, as he seeks revenge for his treatment alongside a colony of hybridized "megabats." For the NiTainos' two other members, V•G1GANTE and Lúz, the energy of la Estrella del Camino reiterates the mythistorical source of empowerment, referencing the vejigante figure above and the "rivers of light" that connect Marisol and her best friend La La/Lúz to Indigenous and Afro-Puerto Rican spirituality. In this multifaceted framework that Miranda-Rodriguez delineates, perhaps this energy could be framed in terms of *aché*—to borrow the equivalent term in Santería for *axé*—once again returning the reader to the waters of the Crossing, its restorative energy, and the many histories in its wakes.

"The ocean has been portrayed as the earth's last frontier of wilderness, which, in terms of American mythology, positions it as the place for narratives of domination," Alaimo observes (2014, 193). The comics examined here contest the very premise of this statement, refuting hydrocolonialist conceptions of the sea as "wilderness" and *aqua nullius* to read its many wakes and recognize the ocean as a multispecies site of *konesans* and *conocimiento*. These ritual, ancestral, and historical knowledges and practices constitute sea ontologies replete with both memorialization and future potential, exemplifying the reciprocal and sustainable relationships with the environment on which the Afrodiasporic and Indigenous spiritualities presented here are founded. As the reader witnesses, the vengeful response to the threat of extinction by the hornets, bees, and wasps in *Is'nana the Were-Spider* and the decimation of bird, frog, and bat populations referenced in *La Borinqueña* are countered by their protagonists who embody this reciprocity. Viewing this relationship in *Marassa* through Strongman's

(2019) and Alaimo's (2016, 133) lenses of transcorporeality intertwines the spiritual with the "material, discursive, economic, ecological, and biopolitical" in an oceanic framework that resonates with the other comics examined here. The necessity for action beyond these comics is clear, as Miranda-Rodriguez has demonstrated: *La Borinqueña Guest Starring Rosario Dawson* (Miranda-Rodriguez et al. 2022) was produced in collaboration with the nonprofit environmental advocacy group National Resources Defense Council, and the *Ricanstruction* (Miranda-Rodriguez 2018) anthology accompanied the creation of the La Borinqueña Grants Program, which supports various local organizations and projects, including those focused on environmental justice and sustainable farming, in its vision for "a Puerto Rico that is self-sufficient and thriving" (Somos Arte, n.d.). Similarly, the decolonial narrative of reciprocity and sustainability found in the spiritual and mythological crosscurrents of these comics seeks to reject further (neo)colonial actions, presenting worldviews that have long contested the premise of the extractivist, hydrocolonial relationship with the environment that has led to the current crises we face. They do not present solutions to be mined for their efficacy on a broader scale but rather descend *anba dlo*, "beyond space-time, beyond self," to weave transoceanic tapestries of African, Afrodiasporic, and Indigenous spiritualities and write existing and new histories, stories, mythologies, and epistemologies into the "third space" of resolution, the mythistorical archive of the waters (Tinsley 2018, 146). In so doing, they reorient readers toward mutually sustainable and reciprocal relationships between the human, nonhuman, and more-than-human, which here are rooted in the multiple and intersecting sea ontologies of the Caribbean and the wider region.

Notes

1 Note: all the comics discussed are unpaginated, which is reflected in their corresponding references.
2 See Humphrey (2019) for a more in-depth analysis of this topic in reference to the Santerians and Brother Voodoo/Doctor Voodoo.
3 The *lwa* in Haitian Vodou are not gods but rather are more accurately described as spirits. As in Santería and Candomblé, Vodou has a single Supreme Being who is notionally equivalent to a god; the *lwa, orishas*, and *orixás* (in Vodou, Santería, and Candomblé, respectively) are the spirits who interact with practitioners.
4 Importantly, as Alexander writes: "Housed in the memory of those enslaved, yet not circumscribed by it, these Sacred energies made the Crossing. But they did not require the Crossing in order to express beingness. They required embodied beings and all things to come into sentience, but they did not require the Crossing" (2005, 292).
5 The second volume of *Mulatako* was created in collaboration Carine Bahanag (Dibussi and Bahanag 2020).
6 All translations from Portuguese are by Sarah Neubecker, student research assistant from the class of 2022 at Colgate University, NY. This quote comes from the volume's back cover.
7 For further discussion of *axé* and Candomblé, see Murrell (2010, 159–182), Matory (2005), and Afolabi (2005).
8 The white bow tied around trees here symbolizes Ossaim, *orixá* of sacred leaves and healing; leaves are a further repository of *axé*.

9 Obatalá, also called Oxalá, is often considered the father of the *orixás* and the *orixá* of love and peace (see, for example, Matory [2005, 140–143] and Voeks [2003, 57–59]). Reginaldo Prandi (2001, 23–24)—whom Canuto (2018) also cites—describes Oxalá as the creator of humankind, with the associated *orixá* Odudua as creator of the earth.
10 The city is named both Cidade Mãe and Oxogbô (Portuguese spelling of Ọṣogbo) in the comic. As Canuto (2021) notes, Cidade Mãe is a reference to the city of "Salvador and the importance of Oshun in Bahian culture and beliefs."
11 The most common color associated with Oxum is gold, as Murrell (2010, 174), Voeks (2003, 59), and Prandi (2001, 22) all observe.
12 The Gede are Vodou spirits who reside in the world of the dead; they are associated with sex, death, and humor.
13 Initiation *anba dlo* specifically references initiation as a devotee of Lasirenn, *lwa* of the depths of the ocean who is most often depicted as a mermaid. She is a counterpart to the Yoruba *òrìṣà* Yemọja and West and Central African spirit Mami Wata (Bellegarde-Smith 2005, 243; Otero and Falola 2013, xix). In Vodou, Lasirenn is associated with the Ezili family of *lwa*; see Szeles (2011), Tinsley (2018, 16–20), and Brown (2010, 223–225).
14 Some avatars of these spirits are also associated with death, such as Ochún Awé (Cabrera 1974, 71; Bolívar Aróstegui 1990, 118) and Ezili Je Wouj, Ezili Nwa Kè, and Marinèt Bwa Chèch (Dayan 1995, 106, 116).
15 Strongman outlines in his introduction the scholarship about the "constituents of the self" and the varying interpretations of the *gwobonanj* (2019, 12–17). As he concludes, "We can summarize the roles of the two most important aspects of the self by saying that the gwobonanj is consciousness, while the tibonanj is objectivity" (13).
16 For further discussion of the importance of trees and wood in Haitian Vodou and the effects of environmental degradation on African-Caribbean spiritual practice, especially Vodou, see Paravisini-Gebert (2005, 2016).
17 See Alaimo (2010) for extended discussion of this concept.
18 In contrast to volume one, the gutters in the two chapters that continue the central narrative in volume two indicate the plane on which the action takes place: white is the spiritual realm of the Mother Kingdom; black is the physical realm inhabited by humans.
19 The diacritics carried by "gèp" and "myèl" in standard Kreyòl are not used in the comic.
20 Turning to speculative fiction, the Impundulu figure appears as a central character in Jamaican author Marlon James's *The Dark Star* trilogy (2019, 2022).
21 In volume one, Legba's dog is introduced as "Cerberus," a reference to the three-headed dog stationed at the gates to Hades from Greek mythology. However, in volume two, the spelling changes to "Cerebus." Here, I maintain the spelling from volume one and the corresponding connotations this carries.
22 Since 2019, Koeia has been the name of the star HIP 12961, which was proposed by Puerto Rico for the IAU100 NameExoWorlds (2019) campaign.

References

Afolabi, Niyi. 2005. "Axé: Invocation of Candomblé and Afro-Brazilian Gods in Brazilian Cultural Production." In *Fragments of Bone: Neo-African Religions in a New World*, edited by Patrick Bellegarde-Smith, 108–123. Urbana: University of Illinois Press.

Alaimo, Stacy. 2010. *Bodily Natures: Science, Environment, and the Material Self.* Bloomington: Indiana University Press.

Alaimo, Stacy. 2014. "Feminist Science Studies and Ecocriticism: Aesthetics and Entanglement in the Deep Sea." In *The Oxford Handbook of Ecocriticism*, edited by Greg Garrard, 188–204. New York: Oxford University Press.

Alaimo, Stacy. 2016. *Exposed: Environmental Politics and Pleasures in Posthuman Times*. Minneapolis: University of Minnesota Press.

Aldama, Frederick Luis. 2017. "ReDrawing of Narrative Boundaries: An Introduction." *Studies in Twentieth and Twenty-First Century Literature* 42, no. 1: 1–6. https://doi.org/10.4148/2334-4415.1965.

Alexander, M. Jacqui. 2005. *Pedagogies of Crossing: Meditations on Feminism, Sexual Politics, Memory, and the Sacred*. Durham, NC: Duke University Press.

Anderson, Jiba Molei. 2002–. *The Horsemen*. Chicago: Griot Enterprises.

Anderson-Elysée, Greg, Walter Ostlie, and Lee Milewski. 2016. *Is'nana the Were-Spider: Vol. 1*. New York: Webway Comics.

Anderson-Elysée, Greg, Antonello Cosentino, and Francesco Montalbano. 2018a. *Marassa #1*. n.p.: Evoluzione.

Anderson-Elysée, Greg, Daryl Toh, and Walter Ostlie. 2018b. *Is'nana the Were-Spider: Vol. 2*. New York: Webway Comics.

Anderson-Elysée, Greg, Antonello Cosentino, and Francesco Montalbano. 2019. *Marassa #2*. n.p.: Evoluzione.

Anderson-Elysée, Greg, Miguel Blanco, and Angael Davis-Cooper. [2021a]. *Is'nana the Were-Spider: Showtime*. New York: Webway Comics.

Anderson-Elysée, Greg, David Brame, and Justin Birch. [2021b]. "Soul Eaters." In *Modern Mythology: Folklore Retold*, edited by Caleb Palmquist. n.p.

Anderson-Elysée, Greg, and Keithan Jones. 2022. *Is'nana the Were-Spider and the Daughters of the Goddess*. n.p.

Bellegarde-Smith, Patrick, ed. 2005. *Fragments of Bone: Neo-African Religions in a New World*. Chicago: University of Illinois Press.

Benedicty-Kokken, Alessandra. 2015. *Spirit Possession in French, Haitian, and Vodou Thought: An Intellectual History*. Lanham, MD: Lexington Books.

BlackSci-Fi.com. 2019. "A Chat With 'Marassa' Creator/Writer Greg Anderson Elysée." Accessed October 31, 2022. https://blacksci-fi.com/a-chat-with-marassa-creator-writer-greg-anderson-elysee/.

Bolívar Aróstegui, Natalia. 1990. *Los orishas en Cuba*. La Habana: Ediciones Unión.

Brathwaite, Edward Kamau. 1983. "Caribbean Culture: Two Paradigms." In *Missile and Capsule*, edited by Jürgen Martini, 9–54. Bremen: University of Bremen.

Brown, Karen McCarthy. 2010. *Mama Lola: A Vodou Priestess in Brooklyn*. 3rd ed. Berkeley: University of California Press.

Bystrom, Kerry, and Isabel Hofmeyr. 2017. "Oceanic Routes: (Post-it) Notes on Hydro-Colonialism." *Comparative Literature* 69, no. 1 (April): 1–6. https://doi.org/10.1215/00104124-3794549.

Cabrera, Lydia. 1974. *Yemayá y Ochún: Kariocha, Iyalorichas y Olorichas*. Miami: C&R.

Canuto, Hugo. [2017]. *Contos dos Orixás: Artbook*. Vol. 1. Salvador, Brazil: Hugo Canuto.

Canuto, Hugo. 2018. *Contos dos Orixás*. Salvador, Brazil: Hugo Canuto.

Canuto, Hugo. 2021. "*The Making of Tales of the Orishas with Hugo Canuto*." Global Comics Lecture Series, Ohio State University, Columbus, April 9.

Chute, Hillary. 2016. *Disaster Drawn: Visual Witness, Comics, and Documentary Form*. Cambridge, MA: The Belknap Press of Harvard University Press.

Coates, Ta-Nehisi, Wilfredo Torres, and Chris Sprouse. 2016–2018. *Black Panther: Avengers of the New World*. New York: Marvel Comics.

Dayan, Colin (Joan). 1995. *Haiti, History, and the Gods*. Berkeley: University of California Press.

DeLoughrey, Elizabeth M. 2007. *Routes and Roots: Navigating Caribbean and Pacific Island Literatures*. Honolulu: University of Hawai'i Press.

DeLoughrey, Elizabeth M. 2017. "Submarine Futures of the Anthropocene." *Comparative Literature* 69, no. 1: 32–44. https://doi.org/10.1215/00104124-3794589.

DeLoughrey, Elizabeth M., and George B. Handley, eds. 2011. *Postcolonial Ecologies: Literatures of the Environment*. New York: Oxford University Press.

Dibussi, Reine. 2017. *Mulatako #1*. [Montigny-sur-Loing?]: Afiri Studio.

Dibussi, Reine, and Carine Bahanag. 2020. *Mulatako #2*. [Montigny-sur-Loing?]: Afiri Studio.

Fernández Olmos, Marguerite, and Lizabeth Paravisini-Gebert. 2022. *Creole Religions of the Caribbean: An Introduction from Vodou and Santería to Obeah and Espiritismo*. 3rd ed. New York: New York University Press.

García, Enrique. 2020. "*Turey El Taíno* and *La Borinqueña*: Puerto Rican Nationalist and Ethnic Resistance in Puerto Rican Comics Dealing with Taíno Cultural Heritage." In *Graphic Indigeneity: Comics in the Americas and Australasia*, edited by Frederick Luis Aldama, 210–233. Jackson: University Press of Mississippi.

García, Ivonne M. 2018. "Diasporic Intersectionality: Colonial History and Puerto Rican Hero Narratives in *21: The Story of Roberto Clemente* and *La Borinqueña*." In *The Routledge Companion to Gender, Sex and Latin American Culture*, edited by Frederick Luis Aldama, 71–82. New York: Routledge.

Gil, Joamette, ed. 2021. *Mañana: Latinx Comics from the 25th Century*. Portland, OR: P&M Press.

Hofmeyr, Isabel. 2022. *Dockside Reading: Hydrocolonialism and the Custom House*. Durham, NC: Duke University Press.

Hollman, Deirdre Lynn. [2022]. "Is'nana and Friends: Uniting Stories and Knowledge from across the Black Atlantic." In *Is'nana the Were-Spider: Birthday Day*, by Greg Anderson-Elysée and George Gant. New York: Webway Comics.

Hopkinson, Nalo, Dan Watters, and Dominike Stanton. 2018–2020. *House of Whispers*. Burbank, CA: DC Comics.

Humphrey, Paul. 2019. "'Yo soy Groot': Afro-Caribbean Religions and Transnational Identity in the Comic Metropolis." *Studies in Comics* 10, no. 1 (July): 115–134. https://doi.org/10.1386/stic.10.1.115_1.

Humphrey, Paul. 2021. "Framing a Decolonial Future: Hurricane María in Independent Puerto Rican Comics." *Latin American Literary Review* 48, no. 96 (Summer): 61–74. https://doi.org/10.26824/lalr.238.

James, Marlon. 2019. *Black Leopard, Red Wolf*. New York: Riverhead Books.

James, Marlon. 2022. *Moon Witch, Spider King*. New York: Riverhead Books.

Jennings, John. [2019]. "Ringing True: Ballads, Rhymes, and the Creation of History." In *Is'nana the Were-Spider: The Ballads of Rawhead & John Henry*, by Greg Anderson-Elysée, Walter Ostlie, and David Brame. New York: Webway Comics.

Lee, Stan, Jack Kirby, and George Roussos. 1964. *The Avengers #4*. New York: Marvel Comics.

Matory, J. Lorand. 2005. *Black Atlantic Religion: Tradition, Transnationalism, and Matriarchy in the Afro-Brazilian Candomblé*. Princeton, NJ: Princeton University Press.

McCloud, Scott. 1993. *Understanding Comics: The Invisible Art*. New York: HarperCollins.

McNeal, Keith. 2010. "Pantheons as Mythistorical Archives: Pantheonization and Remodeled Iconographies in Two Southern Caribbean Possession Religions." In *Activating the Past: History and Memory in the Black Atlantic World*, edited by Andrew Apter and Lauren Derby, 185–244. Newcastle-Upon-Tyne: Cambridge Scholars.

Miranda-Rodriguez, Edgardo, ed. 2018. *Ricanstruction: Reminiscing and Rebuilding Puerto Rico*. New York: Somos Arte.

Miranda-Rodriguez, Edgardo, Emilio Lopez, and Will Rosado. 2016. *La Borinqueña #1*. New York: Somos Arte.

Miranda-Rodriguez, Edgardo, Manuel Preitano, Will Rosado, and Christopher Sotomayor. 2018. *La Borinqueña #2*. New York: Somos Arte.

Miranda-Rodriguez, Edgardo, Will Rosado, and Christopher Sotomayor. 2021. *La Borinqueña #3*. New York: Somos Arte.

Miranda-Rodriguez, Edgardo, Will Rosado, and Christopher Sotomayor. 2022. *La Borinqueña Guest Starring Rosario Dawson*. New York: Somos Arte.

Molina, Jose, Simone Bianchi, and Israel Silva. 2016. *The Amazing Spider-Man: Amazing Grace*. New York: Marvel Comics.

Murrell, Nathaniel Samuel. 2010. *Afro-Caribbean Religions: An Introduction to their Historical, Cultural, and Sacred Traditions*. Philadelphia: Temple University Press.

NameExoWorlds. 2019. "IAU100 NameExoWorlds Approved Names." Office for Astronomy Outreach. Accessed October 31, 2022. https://www.nameexoworlds.iau.org/_files/ugd/6358ac_5eebee4eba4f41b7a9f6201123673a24.pdf.

Otero, Solimar, and Toyin Falola, eds. 2013. *Yemoja: Gender, Sexuality, and Creativity in the Latina/o and Afro-Atlantic Diasporas*. Albany: State University of New York Press.

Pané, Ramón. 1999. *An Account of the Antiquities of the Indians*, edited by José Juan Arrom, translated by Susan C. Griswold. Durham, NC: Duke University Press.

Paravisini-Gebert, Lizabeth. 2005. "'He of the Trees': Nature, the Environment, and Creole Religiosities in Caribbean Literature." In *Caribbean Literatures and the Environment: Between Nature and Culture*, edited by Elizabeth DeLoughrey, Renée K. Gosson, and George B. Handley, 182–196. Charlottesville: University Press of Virginia.

Paravisini-Gebert, Lizabeth. 2016. "*Gade nan mizè-a m tonbe*: Vodou, the 2010 Earthquake, and Haiti's Environmental Catastrophe." In *The Caribbean: Aesthetics, World-Ecology, Politics*, edited by Chris Campbell and Michael Niblett, 63–77. Liverpool: Liverpool University Press.

Prandi, Reginaldo. 2001. *Mitologia dos Orixás*. São Paulo: Companhia das Letras.

Pressley-Sanon, Toni. 2013. "One Plus One Equals Three: Marasa Consciousness, the Lwa, and Three Stories." *Research in African Literatures* 44, no. 3 (Fall): 118–137.

Pressley-Sanon, Toni. 2017. *Istwa across the Water: Haitian History, Memory, and the Cultural Imagination*. Gainesville: University Press of Florida.

Quesada, Joe, Danny Miki, and Richard Isanove. 2005. *Daredevil: Father #2*. New York: Marvel Comics.

Sharpe, Christina. 2016. *In the Wake: On Blackness and Being*. Durham, NC: Duke University Press.

Shelton, Allison Nowak. 2019. "Learning from Rivers: Toward a Relational View of the Anthropocene." *English Language Notes* 57, no. 1 (April): 152–159. https://doi.org/10.1215/00138282-7309755.

Somos Arte. n.d. "La Borinqueña Grants Program." La Borinqueña. Accessed October 31, 2022. https://www.la-borinquena.com/la-borinquena-grants-program.

Straczynski, J. Michael, John Romita, Jr., and Scott Hanna. 2003. *The Amazing Spider-Man: Vol. 4. The Life and Death of Spiders.* New York: Marvel Comics.
Strongman, Roberto. 2019. *Queering Black Atlantic Religions: Transcorporeality in Candomblé, Santería, and Vodou.* Durham, NC: Duke University Press.
Szeles, Ursula. 2011. "Sea Secret Rising: The Lwa Lasirenn in Haitian Vodou." *Journal of Haitian Studies* 17, no. 1 (Spring): 193–210. https://www.jstor.org/stable/41711916.
Tinsley, Omise'eke Natasha. 2018. *Ezili's Mirrors: Imagining Black Queer Genders.* Durham, NC: Duke University Press.
Voeks, Robert A. 2003. *Sacred Leaves of Candomblé: African Magic, Medicine, and Religion in Brazil.* Austin: University of Texas Press.
Walcott, Derek. (1979) 1986. "The Sea Is History." In *Derek Walcott: Collected Poems 1948–1984.* New York: Farrar, Straus & Giroux.
Walford, Jerome, ed. 2020. *Gwan: Volume 2.* New York: Forward Comix.

PART III

Artistic practices, myth, and environmental resilience

6

"GIANT BY THINE OWN NATURE"

Jean-Baptiste Débret and Antônio Parreiras' mythic Brazilian land(scape)s through a transatlantic gaze

Esther Lezra and Esther Sánchez-Pardo

Introduction

In this chapter, reflecting on the impact of colonialism in the Americas and on the quick appearance, in the aftermath of the wars for independence of emergent republics from the United States to south of Argentina, we will focus on the transition of the Luso-Brazilian Empire from monarchy to republic. In such a rich, heterogeneous context, we will address two major issues as they appear in the work of two crucial artists, of French and Brazilian origin, respectively Jean Baptiste Débret (1768–1848) and Antônio Parreiras (1860–1937), whose trajectory is inextricably linked to Brazil as an exotic, wealthy, and Edenic land. Under a postcolonial lens and through a reading of landscape as an instrument of power and domination, our questions revolve around the reasons why the mythologization of life in the tropics produces alternative modernities and interrogates why institutions and practices erected by the colonial machinery promote ways of seeing which distort land and nature. Within these modernities, the function of myth, embedded as it is in the narration of origins, at the dawn of the modern Republic, offers a solid vantage point to approach backbone elements for life in Brazil since the beginning of the 20th century (Schwarcz 2006). Both painters revisit myths of the land as they appear along the nature–culture divide. In our selection, Débret focuses more acutely on the "naturalization" of social relations across race and taxonomies of domination. Parreiras goes deep into the natural distinctiveness of the immense Brazilian land (biological, ecosystemic interactions), which ends up being awe-inspiring. Both of them manage to fuse the mythical–natural and the historical in their canvases in order to teach Brazilian and foreign audiences more about Brazil.

Homogenous, empty time is at the origin of the modern idea of the nation and of nationalism, according to Benedict Anderson (2006). When the time

DOI: 10.4324/9781003348535-10

that frames the modern colonial nation is set in motion and moves around a series of points of convergence where people come together, coercion, domination, control, slavery, the exploitation of labor, social stratification, and structural violence are almost inadvertently well underway. When the imagined community of the modern nation is understood as structured, cohesive, unwavering, and projected into the future, a technology of practices (map, census, Academy, museum) is set to measure, quantify, organize, calibrate, and assess. Territoriality, population, ethnic stock, natural resources, shared traditions, rituals, and memory are subject to scrutiny. How does an artist, in full exercise of her freedom, contribute to the making of the modern nation? Between mythical narratives of origins, cohesive storytelling, familiar images of the land and its peoples, and a distinctive environmental milieu, artists fully engage in the construction of the nation.

Myths of the land

Occupying a pivotal place on Europe's path to modernity, Portugal retained the hegemony of exploration, navigation, trade, and early expansion together with Spain and followed by other Western European countries. Ever since the Portuguese explorers set foot on Brazil, allusions to the Garden of Eden are found in early documents and letters where the wonders of Brazilian nature are described.[1] This terrestrial Paradise in which humans, luxuriant nature, wild and exotic animals coexisted, surrounded by bountiful and inexhaustible resources in a pleasurable state of innocence, found its way from religious accounts to art and literature. In the late Middle Ages, the European idea of Brazil was immersed in such a context, which, according to historian Murilo de Carvalho's account, showed "God's generosity regarding the country [being] so great that, besides the gift of natural paradise, he also promised historic paradise. The greatness of nature was the promise of a future big and powerful empire" (as cited in Eckl 2015, 220). In Carvalho's view, the myth of the great future empire elicits a desire for national greatness, for a status of power, which is not backed by the necessary efforts to turn that dream into reality. The reputed historian argues that "finally, paradise is destroyed and the empire does not materialize" (2001, 112).

This idea of a terrestrial paradise grounded the Brazilian national myth in an exuberant Nature with overabundance and inexhaustible resources up until the first quarter of the 20th century and, from its inception, catered to the needs of the authoritarian regime of Getúlio Vargas (1930–1945). It was at this point that Brazil looked into its past and engaged in constructing the country's collective memory. The persistence of this myth of paradisial dimensions has been a constant within the national imaginary.[2] In the 20th century, this land of boundless opportunity and riches ready to be plucked by the intrepid and ambitious found its way into politics, especially in the country's expansion to the interior (construction of the new capital of Brasilia in the 1960s) and the Amazonian wilderness. Brazil's power and leadership founded its promising future on the expansion of its tropical frontier.

Major Brazilian symbols like the flag or the national anthem are also embedded within the context of an Edenic land where one can fulfill human longings and expectations. As the Brazilian anthem goes, "Giant by thine own nature, / Thou art beautiful, strong, a fearless colossus, / And thy future mirrors that greatness." The lines just quoted come from poet Antônio Gonçalves Días, "Canção do Exilio" (1998, 105). From his "exile" in Coimbra—he felt exiled in Portugal—he sang the beauties of his Brazilian homeland.[3] This anthem, composed in two periods, music and lyrics (1831 and 1922, respectively), was a product of the Romantic idealization of Brazil as homeland, following the Parnassian style and rhetoric that is meant to invest the country with an exotic atmosphere and undaunted new world aesthetic.

Going back in time, from Pedro Alvares de Cabral's discovery of Brazil and the account provided by Vaz de Caminha's letter to King Manuel I of Portugal, we get the first page of Brazilian history. In the letter, the rich description of the land goes side by side with the portrayal of its indigenous peoples. Caminha writes,

> We caught sight of men walking on the beaches. The small ships that arrived first said that they had seen some seven or eight of them. We let down the longboats and the skiffs. The captains of the other ships came straight to this flagship, where they had speech with the admiral. He sent Nicolau Coelho on shore to examine the river. As soon as the latter began to approach it, men came out on to the beach in groups of twos and threes, so that, when the longboat reached the river mouth, there were eighteen or twenty waiting. [...] They have good, well-made faces and noses. They go naked, with no sort of covering.
>
> *(2019, 12)*

Caminha described the complexion of the peoples he encountered as being "dark brown, rather reddish color" (2019, 12) and stressed the lack of social organization and religion in their ways of living, noting that "these people are good and have a fine simplicity. Any stamp we wish may be easily printed on them [...]" (13). The natives' features are depicted in Caminha's account in great detail, both men's and women's: how they painted themselves, wore their hair, the shape of their noses, and their facial features, including their nudity and lack of body hair. The letter reveals how these new territories offered lush, semitropical lands populated by peaceful peoples whose ignorance and primitivism, together with their strength and healthy habits, could well be a great asset to the Portuguese Crown and thus worth exploring in depth.

Claude Hulet, researching Caminha's letter as the first testimony of Brazil's entry into Western history, points out the merchant's emphasis on two major factors relevant for our argument:

> 3) the fertility, beauty and the abundance of the new land comparing it to regions of Portugal; and, finally, 4) the natives of the newly discovered land, whom Caminha described in considerable physical and moral detail, emphasizing their fine appearance and excellent health, their Edenic innocence and skittishness, and their eager generosity and readiness to be of service.
>
> *(1995, 123–124)*

Hulet skillfully captures the merchant's first impressions of the land. Vaz de Caminha's is a full description of objective elements, which allow us to catch a glimpse of the landscape he observed when he first saw Brazil.

US poet Elizabeth Bishop (1911–1979), lusophile par excellence, who spent 16 years of her life in Brazil (1951–1967), argues that in Caminha's letter there is probably the first strand of Romantic indigenism, centered upon the myth of the noble savage which would later so vigorously permeate the imagination of Brazilians (2011). After much historical and ethnographic research and a series of debates on the opposition and insurmountable barrier attributed to barbarian man versus civilized western man, it is undeniable that Montaigne holds the "official" position on the topic of the noble savage. In his essay "Des cannibales" (Montaigne 1993), the French philosopher demonstrates through his rhetoric and tropes that the Portuguese are "barbarous" as opposed to the cannibales of Brazil who, in reality, are more "civilized." Privileging an account based on experience, as Montaigne claims—on eyewitness accounts and on his own conversations with them[4]—it is clear that he used his sources liberally to depict the cannibals with a positive twist. When contrasting French society with that of their uncivilized counterparts, he overestimates the state of nature in which they lived. Among them, there is no commerce, no social inequality, no inheritance, no public administration, and they do not know envy, avarice, lying, and betrayal:

> They do not strive for conquest of new territories: since their own still possess such natural fertility as to yield them all their necessities without labour or trouble, in such abundance that they have no need to extend their borders.
>
> *(Montaigne 1993, 114)*

Finally, he is apprehensive of what might become of the noble savage who comes in contact with civilized man. Montaigne builds on a utopian vision of the "au delà" of western civilization.

Myths of the Brazilian land and its peoples coming from colonial contact and exchange go beyond description, calling for a complex process of acculturation with long-lasting consequences for those under colonial rule. Much has been written about appropriation when writing about or facing the Other in postcolonial studies. Nevertheless, mythic representation has not been systematically analyzed with regard to the epistemological grappling with human difference that is the visual-cum-textual rendering of the Other. In this chapter, we focus on two instances of the mythic representation of the Brazilian land and its autochthonous population by Jean-Baptiste Débret and Antônio Parreiras in two crucial moments within the history of Brazil, from Portuguese imperial rule to the country's awakening as a new Republic. The history and evolution of the mythic representation of the land and the native Other offers an ideal object of analysis because the Portuguese (and by extension, the French, as we'll argue) cultural imaginary of the native resounds with fantasy and horror. The native land and the indigenous populations are the unknown, inspiring curiosity and fear. In our approach to this Other of the land, this chapter aims to uncover how visual artists imbued the land and Indigenous peoples with their capacity to conjure wonder or anxiety as well, as Débret's and Parreiras' own canvases and writings prove.

Indigenous and Portuguese settlers became protagonists of the Romantic theme. Brazilian literature with Romantic overtones produced artists such as the poet Gonçalves Dias (1823–1864) and the novelist José de Alencar (1827–1877). Indigenous populations, in the midst of tensions of the independence movements in the Americas, antagonized the European. In the field of visual arts, a new generation of painters with Victor Meirelles (1832–1903), Pedro Americo (1843–1905), Angelo Agostini (1843–1910), and Almeida Júnior (1850–1899) created a body of work in which the Brazilian nation hovers in permanent oscillation between more classical models and postromantic derivations. Based on the principles of academic and naturalistic painting and on performative practices of iteration, failure, de-formation, and adaptation, we turn to the works of Débret and Parreiras, who represented and admired Brazil in distinctive yet to a great extent, complementary ways.

Why landscape? Tropicalization and its discontents

Reflecting upon the importance of landscape painting in Western and world cultures, W.J.T. Mitchell sides with a "skeptical reading of landscape aesthetics" ([1994] 2002, 6) and argues along the lines of Michel de Certeau and Henri Lefebvre that "the expression of power in the landscape [is] a manifestation of law, prohibition, regulation and control" (x). Mitchell's reading of K. Clark's (1949) views ironically leads him to argue that "the 'reflective' and imaginary projection of moods into landscape [can be] read as the dream work of ideology [while] the 'rise and development' of landscape [can be] read as a symptom of the rise and development of capitalism [...]" ([1994] 2002, 7). Mitchell calls for landscape to be understood "as a process by which social and subjective identities are formed" (1).

It is our contention that complex forms of both compliance and resistance are found when one focuses on visual representations of nature, peoples, and localities across the Atlantic with Portugal as the cornerstone of empire. In our attempt to disentangle how mythical narratives and environmental concerns come together in the work of Jean-Baptiste Débret and Antônio Parreiras, we will focus on the centrality of nature and landscape—at times a social landscape—as constitutive of life in Brazil and on the mythical narratives that both precede and exceed the central events of the time span under scrutiny, roughly 1816 to 1937—from Débret's arrival in Brazil to the passing of Parreiras.

Mary Louise Pratt has described the cultural struggle waged in South America from the second half of the 19th century onwards as a struggle for modernity, because "the diffusionist momentum of modernity becomes a powerful determinant of reality in all of its dimensions; its empirical particularities are very consequential. This is a truism, but it is one to which metropolitan theorizing on modernity remains remarkably immune" (2002, 38). At the same time, the population of the Brazilian capital and the largest urban centers in the country adapted to the major changes enabled by telegraphy, railroads, industrialization, and transatlantic newspapers. Brazilian intellectuals and the bourgeoisie engaged in debates around how to represent the diversity and richness of the Brazilian nation both at home and abroad.

Srinivas Aravamudan aptly asserts that tropicalization has been both a trope and has featured the conditions of existence of the "tropicopolitans."[5] In his insightful analysis, Aravamudan speaks of a series of tropological recursions. These range from ethnic–linguistic particularization, to generalized condition, to subsequent historical and geographic deduction of those in that condition, to metaphorical application in a different national context (i. e., neighboring countries) allowing "indigeneity" to trope or turn in various contexts (1999, 5). How does one translate into images these tropological operations? How does the land and its dwellers become tropicalized in the aftermath of transatlantic colonial expansion?

A source of historical knowledge, political strategy, cognitive mapping, and onto-epistemological inquiry, inhabiting the West has ever been a self-justifying, self-explanatory operation. Inhabiting "newly encountered" geographies has given rise to tactics and procedures leading to the expansion of Empire and the domination of subaltern zones, geographical areas, and far-away lands. In the following pages, the Portuguese–French metropolis's tropicalization of the Brazilian nation—properly speaking, Brazil lies between the tropics, and its northern section is considered to be under "tropical" conditions—will be discussed by approaching the pictorial work of Débret and Parreiras. This chapter will attempt to show how the visual representation of the Brazilian land retains a mythical dimension that distorts nature and population at the same time that it produces an alternative modernity. How does the individual (un)learn to be modern in the tropics?

In the last decades of the 19th century, Brazilian intellectuals, frustrated with the current situation in the country, intensified the questioning about what Brazil was and what Brazilian national identity consisted of. It was only in the context of the consolidation of the Republic that the debate emerged within visual representation. The local elites commissioned official testimonies that produced meanings to mark history with signs of their power and the newly remodeled social order. As scholars of the period hold, this led to an "official art" (Álvarez 2009; Salgueiro 2009; Cerdera 2012), and the artists and their audiences' reactions can be related to a vast stream of other works—ancient and modern—from all over the world engaged in producing art for the masses in the public sphere.

In Brazil, historical painting dates back to the foundation of the Academia Imperial de Belas Artes, an institution that emerged around 1820 as a result of the French Artistic Mission of 1816.[6] At the moment of the shift of status of the colonies to the United Kingdom of Portugal, Brazil, and the Algarves, it was necessary to reproduce the images of the Lusitanian royalty newly based in Rio de Janeiro in order to gain visibility in the rest of the world, as well as to provide the official symbols for Brazilian society. In 1822, the processes of independence and the emergence of Brazil as a nation demanded the creation of symbols capable of transmitting and establishing the new status. Brazil became a new nation in the making. Historical painting would provide the iconographic symbols of the nation and the Imperial Academy would train future professionals for such a project. Historical painting, which required extensive documentary research to be carried out in depth, urged artists who practiced this genre to become "historians in their own right," as Peter Burke states (2001, 158). Even so, the images produced are subject to interpretation, because they were not restricted to reproducing scenes from the past as they occurred. In a moment of rearrangement of the intellectual field, such as the transitional period from empire to the first years of the new republican regime, in which history itself was in dispute, the artist's role as historian was enhanced.

Débret's Brazilianness

Voyage Pittoresque et Historique au Brésil, ou Séjour d'un Artiste Français au Brésil is an artistic and narrative representation of the land and of the colonial, indigenous, and black peoples and cultures of Brazil authored by the French neo-classical artist Jean-Baptiste Débret between 1816 and 1831.[7] Written in Brazil during its transition into independence from Portugal, this cultural representation follows a Frenchman's observation of the decline of Portuguese rule and the rise of Brazil as an independent political entity. But the dates that mark the beginning and end of the period of documentation are significant beyond marking the decline of Portuguese rule and the rise of Brazil as an independent political entity: Débret begins his expedition to Brazil only a decade after France's loss of Haiti, and France occupies Algeria five years before Débret finalizes his period of documentation. Read in this context, *Voyage Pittoresque* reveals an example of the workings of the French colonial and imperial imaginary as it transitions from an American to a North African-based Empire. Débret's text shows a post-Haitian and pre-Algerian French imaginary struggling to work through and pacify—through vicarious participation in the emergence of the Brazilian Empire and colonial practice, which is described as benign, productive, and largely promising.

Débret's connection with artists and intellectuals of the French bourgeoisie in France and Brazil make this text key to gain perspective on the transnational and trans-spatial itinerary of France's imperial self-imaginings. The Débret family was related to Desmaisons, the royal architect and designer of the façade of the Palais de Justice in Paris, as well as to the renowned artist of the French Revolution, Jacques-Louis David. His father had been a scribe for the jury of Paris, a student of natural history, and friend of Dauberton and Lesage. After the strategic losses in Europe and Egypt that brought Bonaparte's imperial vision to a close, Jean-Baptiste Débret was recruited by the Portuguese monarch João VI, who had previously fled Portugal to Brazil to escape persecution at the invasion of Portugal by the forces of Napoleon I. Débret formed part of a mission of French artists, sculptors, and engineers who were hired by João VI to go to Rio de Janeiro to establish the Academia Imperial de Belas Artes. During his stay there, Débret was charged with training Brazilian artists as well as painting historic scenes commissioned by the monarch. He made extensive visual documentation of the peoples and customs of Brazil, from which emerged his written and artistic documentation of Brazil.

We read his written and visual scenes for what they tell us about Brazil at this historical juncture. The community of French artists arrived in Brazil a few days after the death of the monarch, Joao VI, when the coronation ceremony for the new Portuguese monarch, Pedro, who would later be crowned the first emperor of Brazil, was underway. In his Introduction to Débret's *Voyage*, Raoul Rochette makes very clear that the purpose of the community of French artists was to contribute their talent to the placement of Brazil among the civilized kingdoms of the European continent. He writes,

> It brings great pleasure to us, and glory to France, to think that in 1816 several of our compatriots and distinguished artists, several of whom were members of our Académie, devoted such zeal and courage to go to this land still in its infancy in order to propagate a taste for the arts; it is glorious, let us say, that they have managed, with the force of perseverance, to found an academy of fine arts, and to contribute to make Brazil participate in the great works of our civilization.
>
> *(Rochette [1865] 1965, 7)*

The events that transpired in Brazil, which Débret was to document, would eventually lead to the foundation of the Empire of Brazil, independent from Portugal. Débret's employment and artistic production in the Academia de

Belas Artes in Rio was thus integrally related to the emergence of Brazil as an empire independent of Portugal. Débret's experience documenting political events of Napoleon's successes before his defeat would have made him keenly aware of the role that the production of art played in the shaping and securing of the state imaginary. His art had done this work for the Napoleonic state and was charged with a similar responsibility in Brazil.

As argued before, the perspective through which Débret experienced Brazil had been shaped and informed by the legacy of Napoleon Bonaparte's imperial campaigns and recent failures—Haitian Independence and the failed campaign in Egypt. This can be seen in the sensitivity evident in his descriptions of Brazil to the violence unleashed by colonialism. This sensitivity to, or nervousness about, colonial violence first emerges in the first volume of the *Voyage*, where Débret describes the effect on the mind of the colonizer entering Brazil for the first time into a land whose scenery and peoples are described as untouched by civilization, certainly within the state of nature (Montaigne, Rousseau). Débret's language shows us the mind of a "courageous European" braving the wilds of Brazil in 1816. This "courageous European" is haunted by the figure of the European who had arrived in the "New World" centuries before. Débret's description of his own entry into Brazil doubles as a reconstruction of a generic scene of primal encounter. The painter writes that it was "impossible not to marvel at the miraculous greenery [and that] in the forest, man was obliged to become humble" ([1865] 1965a, 21). Through speculation he identifies with and re-creates the mythical European explorer and wonders "[h]ow many obstacles must be overcome by the courageous European who wants to observe up close the customs of the Brazilian savage [and] [h]ow would it be possible for the traveler not to be filled with respect and admiration at the sight of this miraculous fecundity?" ([1865] 1965a, 21). Speaking from a generalized subject position, Débret asserts that in the Brazilian forest "[o]vercome by the impotence of his physical abilities the man, so brave otherwise, is humbled. [...]" ([1865] 1965a, 21). The "courageous European" is humbled. Yet, it is not only the generic European civilized man who feels humbled by the Brazilian forest. In 1816 it is Débret, a civilized Frenchman carrying with him the recent memory of Napoleonic grandeur and success who is humbled. A man who had born witness and had artistically documented and espoused the rise of Napoleon; who had celebrated Napoleonic successes through his art, who had subsequently seen Napoleon humbled by the loss of Haiti and the failure of the campaign in Egypt and seen him turn his eyes, in the wake of these losses, toward Algeria. Reflected in Débret's consciousness and the mythical consciousness of the primal European explorer is the wider imperial mind. The Brazilian forest humbles the imperial consciousness at the same time that it allows it to generate, imagine, and move toward a new type of colonial discovery and new kind of empire; this fecund material and imaginary space enable a movement away from the memory of an empire tarnished by the repeated failures of the French enterprise in Europe, the Caribbean, and Egypt.

The move to re-create the primal scene of discovery and encounter is also pivotal to the framing of the second volume of the first book, where Débret reintroduces the reader into the mind of the mythical colonizer in a romanticized, almost nostalgic account:

> Practically thrown onto the coast of Brazil, at first the Portuguese man timidly penetrates the forests nearest to the beach, where he builds fortifications. / On the other hand, the indigenous man, afraid of the appearance of an unknown man, observes him from a distance, hidden behind the thick vegetation of his virgin forests. Meanwhile, a secret attraction brings them together and soon the open-heartedness of the indigenous man gives in to the seduction of the European. Gifts and mutual favors establish the first links.
>
> *(Débret [1865] 1965b, 5)*

It is more than a romanticized account of discovery and possibility. It is a reconstruction of an event that the European imaginary (symptomatically present in Débret's narrative) remembers leads to the unleashing of centuries of turmoil and violence. The new narration writes over the violence unleashed by the sustained inhuman mistreatment of human beings in the French Caribbean colonies: it rewrites the scene of contact in what we see as an attempt to suture a wounded imperial imaginary and to infuse it with renewed vigor. In this re-created scene, the European traveler comes into contact with an indigenous tribe that, only initially hostile, is easily pacified by a few words uttered by the guide, leading to the initiation of a mercantile relationship between the colonizer and indigenous people. The present tense in this re-created scene evokes a mythical magical timelessness of the originary contact, the lost possibility of another kind of colonial sociability and the ongoing hope of a new empire that is unfolding in the French imaginary as Débret writes. But the scene of enchantment (the re-creation of the lost possibility of another less disastrous kind of empire) and the attendant suspension of the memory of violence only last a moment and are exploded by the tragedy that breaks through the momentary illusion. Débret continues,

> mutual gratitude had joined [the colonizer and the "savage man"] almost entirely until the greed of the sovereigns of Europe throws into their midst military forces that destroy, in an instant, several years of social relations.
>
> *([1865] 1965b, 5)*

In these few words, Débret's narrative releases a spasm of disenchantment with empire and the colonial enterprise in general, in particular relation, perhaps, to the memory of the Haitian Revolution and the horrors of a French colonial régime. These spasms of disenchantment recur throughout his text.

The French colonial and imperial imaginary rewrite themselves through Débret with inflections of past imperial violence and black revolutionary agency experienced by France in Haiti, the failure of France to secure Egypt, and the ripening plan to move into North Africa. They do so in the safety of a space in which, at the time, France has significant cultural value (through the ties of the court to the school of fine arts the French artists were commissioned to found) but no direct political responsibility. Relieved of direct political responsibility over this particular empire, the French mind is free to reinvent itself. For a moment, the French colonial imaginary has a chance to rewrite and perhaps even refashion itself as less violent and less savage than it knows itself to be. The repression of the savagery of European policy in the colonies emerges repeatedly in European imaginative cultures of the time, but what is fascinating about this particular instance is that, for the French artist and witness of the horrors of Napoleonic rule, Brazil seems to serve as a mythical land and an enchanted place from which, far from the direct pressures of French politics, to reimagine a better empire as well as a less violent colonizer. The enchanted space allows the struggling French imperial imaginary, marred by a repressed ignominious past, to revive through the vicarious or parasitical experience of empire.

Débret's account of the arrival of the French artists to Brazil offers insight into what his language describes as an enchanted transformation of the Frenchmen as they pass into Brazilian territory. They penetrate this mythical land and enter the space of enchantment in a quasi-magical process:

> The sun hadn't yet emerged on the horizon when all of the artists, already on deck, enthusiastically admired the unique masses of an unknown vegetation whose details disappeared again in the wave of light mist that half-enveloped them.
>
> Eager to examine that precious scene, whose details and tones were absolutely new to us and increasingly seductive as the sun revealed them, we finally discovered the enchanting view of that sumptuous place, covered everywhere by a brilliant dark green shine [...].
>
> *([1865] 1965b, 33)*

The artist is compelled, by what seems to be a force beyond his control, by what he sees:

> Upon the heels of the first emotion of admiration followed the desire to document the scene, and, taking the pencil sharpened the night before I began to carefully trace the outlines of the place we were.
>
> *([1865] 1965b, 33)*

The sense of enchantment reappears after a passage in which Débret describes an offering of delicious fruit brought from the mainland aboard the ship in a scene reminiscent of a scene of temptation, marked by imagery of fruit and the mythical promised land. The translator doubles as a magician: finally, after hours of waiting for permission to land, the French travelers receive word from the Portuguese court; the artists are escorted by a Brazilian translator and they disembark, "already Brazilians, but always French at heart" (Débret [1865] 1965b, 34). The French imaginary is thus, as if by force of enchantment and the mythical allure of the land, transposed into the Brazilian imaginary. From this new space, the French imaginary appears freer to reimagine empire.

Throughout the volume, enchantment mixes with regret and horror at the treatment of the enslaved populations as Débret describes the mistreatment of African people newly arrived in Brazil. Perhaps Haiti haunted Débret's observations of and musings on the lives and labor of the vast enslaved population on whom the entire Brazilian economy rested. The enchanted narrative of early colonial contact is quickly subsumed in regret and horror as Débret observes that as a result of the destruction of the social relations between colonizer and colonized and the treachery and greed of the European colonial enterprise, the indigenous peoples had ceased to engage in commerce, thus obliging Europeans to resort to the importation of African slaves to build and sustain the colonial economy (Débret [1865] 1965b).

Haiti had impressed the imperialist imaginary of France with a lesson about empire. For the French, the failure of colonialism in Haiti was due not to the horrors of enslavement but rather to the disastrous economic results of putting power and the infrastructures of material production into the hands of enslaved people. From this perspective, this had inevitably led to the loss of the most valuable colony in the Northern Hemisphere. This loss remained imprinted on the collective imperial imaginary that surfaces in the text of Débret; a loss so deeply horrifying that it is expressed indirectly, in flashes of enchanted rewriting of empire surfaced from the recesses of traumatic memory.

The nature–culture divide: white exceptionalism vs. barbaric Others

Well before feminist theory, environmentalism and new materialism challenged the ontological divide between nature and culture, between the human and the more-than-human; nature and culture remained in separate spheres.[8] From the turn of the century, modern anthropology proved that small-scale societies usually had a closer relation to nature (Descola and Pálsson 1996). The aforementioned goes hand in hand with a Western-centric vision associated to European colonialism and other forms of imperialism.

As argued in the section "Myths of the land," Portuguese imperial expansion to America documented, from its inception, the encounter with primitive peoples purportedly lacking social organization, religion, and education and exhibiting instead innocence and simplicity. Indigenous peoples living mostly within small rural communities spread out throughout the country. Their subsistence economy as well as their social organization, rituals, and ways of life demanded continuous contact with nature. As an essential part of the imperial enterprise, and apart from what happened with the gradual exploitation of the indigenous people, it is a fact that the Portuguese in Brazil imported more African slaves than any other country in the world.

Slave labor was the driving force of the economy in the sugar plantations, as well as later on in mines (gold and diamonds). A tropicalization of the economy based on the extraction of raw materials, agricultural work (coffee and sugarcane, mostly), cattle ranching, and food production, and a fluid trade between Europe, Brazil, and western Africa (for slaves) featured progress as the landmark of this new world power.

Slaves were also used as domestic servants within family homes in big cities. It is not surprising, therefore, that in Débret's extensive pictorial oeuvre there are many examples in which the work of black slaves or the exploitation of indigenous peoples is confined to the natural domain, where culture has not penetrated yet and human labor is equated to the labor power of beasts.

The reimagining of empire occurs, in Débret, through the objects he chooses to describe in his written and visual narratives. Débret's description of the slaughter of cattle in Rio (see Figure 6.1) provides another example of a narrative that is apparently about Brazil but, when read closely, yields a narrative that is more about what Brazil represents to the French imperial mind, a distorted image of the young nation in the making.

Débret observes "with horror that despite the immense progress of civilization in the capital, some elements of Brazilian barbarism remained [...] in the method of slaughtering cattle," which consists in herding about 40 cattle into an open path in the streets, fortified with fences on either side; once enclosed, two or three black men armed with axes attack and kill them, one after the other ([1865] 1965b, 98). He is disturbed by the violent slaughter but asserts that it was "no less disgusting" ([1865] 1965b, 98) to see these same men, still covered in sweat and blood, quench their thirst with a large glass of aguardiente before renewing the slaughter. The last sentence of this passage indicates another source of trouble for Débret. The trouble with this ritual of slaughter is not so much the violent and bloody spectacle that it offers but the relative ease and efficiency with which the three or four men perform the killings, as well as the social connection and bonding that occur between rounds of slaughter during a moment of brief social engagement with each other. This scene, writes Débret, "constituted a horrible spectacle for the European" that lasted for the length of 15 minutes. In the context of the colonial mind, it is possible that the sight of black men felling one cow after another would trigger a reminder of the strength and skillful manipulation of the machete that black people used in self-protection against colonial cruelty

FIGURE 6.1 Jean Baptiste Débret. *Voyage au Brésil* ([1865] 1965b, 216, plate 28).

and in the pursuit of freedom. But perhaps more interesting, the colonial observer is no less horrified at the sight of these men engaging in conversation with each other and their friends. Not only do these men possess tremendous physical strength but they also create and cultivate social networks with the many others like them whose labor drives the Brazilian economy. Read in this way, the description of the slaughter of the cattle becomes a coded text for the deeply feared repetition of concerted and organized black agency that had produced the Haitian revolution and could produce another challenge to colonial order. This group of men embodies the tropicopolitans ethos, the image and alleged character of these strong and fearless *colossi* (whose strength and resilience are equated to that of the Brazilian nation as in the poem used in the national anthem) stand in stark opposition to the white bourgeoisie who is absent from a social space where labor and slave and working class's practices hold sway. The written description of the slaughter does not translate into a visual representation, as if the more direct visual representation of black people's skilled wielding of weapons would have been excessive. The illustration that accompanies the text does not give a visual documentation of the slaughter, nor of the blacks who perform the slaughter engaging in social interaction in the break between one slaughter and another.

The figures in the illustration that are related to the slaughter are the two figures in the background carrying what must be freshly butchered meat. The placement of these figures in the background visually diminishes the power and strength of the people whose job it is, first, to slaughter and butcher these enormous oxen and then to transport the meat that is to be eaten by the people at Court. The scene, apparently a peaceful one, shows a bay, mountains, and palm trees, a tropical setting where sociality revolves around slavery, animal domestication for consumption, and economic benefit, as well as the proximity of small towns to the natural environment, which facilitates economic development in its early stages. Just below, a second engraving shows slaves engaged in taming oxen using the yoke. The process of taming is a practice that always goes hand in hand with imperial conquest: the domination and conditioning of the land has a parallel in the domination, control, and submission of its peoples.

The sort of agency Débret chooses to document in his artistic representations of black people in tropical Brazil focused primarily on contexts in which they belong in the realm of nature, and they are physically laboring for or subjected to the structures of empire. This labor is evident in Volume 2, where Débret lists several public works (fountains, public markets, public squares, and buildings) in Rio de Janeiro as evidence of the progress of the colonial civilizing mission ([1865] 1965b, 168–169).[9] Through these, and the immensity of each of the works involved, he indirectly registers the centrality of the labor of enslaved people. Earlier on, his descriptions of herb vendors, shoemakers, hunters, street barbers, and builders also register the different forms of labor and social interaction in the emerging Brazilian society. Of particular note are careful descriptions of the various vessels used to perform different tasks used by laborers to carry different things from place to place in order to perform the strenuous labor of building, feeding and sustaining the infrastructure of the Brazilian empire. Toward the end of the volume, Débret further captures the struggle of European civilization with the older societies of Brazil as he reflects,

> One can finally believe, seeing all the preparations underway, that very soon the streets of the city, as well as the road to Saint-Christophe will be illuminated by gas-lamps. / In a word, everything is improving in this country, where the progress of Enlightenment, under which Brazil has emancipated itself, has given it noble distinctions in science, the arts, and luxury. / Would it that this speedy march toward civilisation never alter the well-known friendly hospitality of Brazil, characterized through

> several centuries by its naturally good people, who deserve to figure foremost among the generous nations of which Europe boasts.
>
> *([1865] 1965b, 169–170)*

As it comes through in Débret's reflection, nature and culture, occupying worlds apart notwithstanding, operate on a continuum, because slavery and capitalist expansion feed off each other.

In juxtaposition to the interpretive difficulty that Débret's observation of black people's spontaneous public gatherings shows, some of his observations of society in Brazil show a capacity of sharp insight into the complex ways that cultural practices could signify.[10] A few pages into the second book there is a particularly fascinating example of Débret's vision when he has a full understanding of what he sees. The painter opens the second book of his *Voyage Pittoresque* with an overview of the political and religious history of Brazil as continually influenced by Europe since 1789. In the first pages of the second book, he forcefully reasserts what he stated at the beginning of the first volume and what has been a recurrent theme throughout the narrative, namely, the key role French artists played in the development and emergence of the great and enlightened empire of Brazil. The insistent and, as we see it, healing repetition of this theme suggests the need—in the author as well as the readers in France to whom his narrative is addressed—to celebrate the influence of France through the dissemination of its culture at a moment when its political influence was on hold. In this context, Débret reflects on the situation of public education as it relates to the foundation of the state. He remarks on the noteworthy strides in the education of women in 1819 and offers an analysis of the reasons society had previously barred women from a public education.

As Débret understands the Brazilian society in Rio, fathers and husbands had preferred that their women be kept illiterate and uneducated to destroy extramarital amorous correspondence. He explains that this enforced illiteracy led women to develop a culture of symbolic communication that resulted in the creation of a language whose grammatical units were composed of different types of flowers and colors. In this language that was handed down through the generations, a flower could express a thought or transmit a request, which could be expanded with the addition of other flowers or leaves taken from herbal plants whose meaning had been previously arranged. The system of communication Débret outlines above stands, as we see it, as a metaphor for the "strange" scenes that catch Débret's attention and yet escape from his grasp. The variety of scenes and objects that make their way into Débret's narrative could be understood to signify, like the flowers, petals, and different combinations of colors, information that circulated in networks and ways that lay beyond the immediate or easy understanding of the European observer. The language of flowers is not only a metaphor for systems of communication that escape ways of collecting and exchanging information that would have been recognizable to the enlightened European mind but an active example of the ease and fluency with which, when necessary, counter-languages composed of rhythms, shouts, cadences, and musical compositions could have been used to cultivate and circulate counter-colonial sentiment, strategy, and literacy.[11] Once again, it is on the fringe between nature and culture that cross-cultural relations take place. Women, whose lives and ways of relating are closer to nature than to the social world, and men, who, invested with patriarchal power, dominate and control society, co-exist within a (supra)national structure that deprives of education and resources to women and colonized subjects. Débret provides conclusive visual testimony to social relations fraught with inequality and coercion. His visual "chronicles" have been instrumental in documenting history and social life in Brazil.

Parreiras and Brazilian nature in landscape painting

In the 19th century, landscape painting was channeled either into the romantic pastoral with a peaceful scenery, clear skies, and rural life or to stormy natural scenarios, present since Neoclassicism and exacerbated by Romanticism. The landscape then responded to and was moved by the human drama and, through this connection, had a strong impact on audiences. The problematization of Romantic landscape painting—in the old and new continents—responded to the production of strategies around the looks and pictorial works that integrated the artistic ethos of specific artists.

In Valeria Salgueiro's view, ever since romanticism developed in the country, landscape painting was consolidated into the genres in which the Brazilianness of the country's art could best be reflected. It should render the forest, woods, and other components of the Brazilian native landscape into pictorial expression, evoked in their unparalleled singularity, in the face of other landscapes of the world (Salgueiro 2009).

Parreiras' work was sensitive to the production of a belated romantic sense and reached an almost unparalleled artistic and intellectual level in his time. Crucial in his practice was his endorsement of landscape art practiced from empirical observation (Salgueiro 2009). His canvases' visual rhetoric became representative of the nation as seen and lived by the white settlers, opening up gradually to the indigenous dwellers of the land. Inserting himself in the frameworks of modernity, his art provided a grounded connection to European civilization from the Brazilian tropics.

In the mid-19th century, at a time when Brazilians felt fascinated by French culture and it became a fad among the elites, Antônio Parreiras was born on January 20, 1860, in the city of Niterói. A very early vocation for painting and his ability to observe and to draw pushed him to pursue his career as a painter. He enrolled in 1883 at the Imperial Academy of Fine Arts. However, unhappy with the formal environment, the following year he moved to his hometown, Niterói, joining the group formed around the German painter

Georg Grimm. He then began taking plein air painting classes, becoming a member of the "Grimm Group," a milestone achievement in landscape painting in the world of Brazilian art. This group focused mainly on natural landscapes, which greatly contributed to the renewal of the genre with the inclusion of specific local elements. In 1888, Parreiras traveled to Venice on a study trip, where he attended the Academia di Belle Arti di Venezia. Upon his return to Brazil in 1890, he founded the Plein Air School of Painting in Niterói, Rio de Janeiro. Between 1906 and 1919, he made several trips to Paris, where he established his studio and was appointed delegate of the National Society of Fine Arts. In 1926, widely acclaimed as an artist, he published his autobiography, a volume entitled *History of a Painter Told by Himself*, thanks to which he was admitted three years later and excelled in the inauguration of the Fluminense Fine Arts Hall in the city of Niterói.

In the early years of the 20th century, he developed a new phase in his production with his historical paintings. In most cases, these historical paintings were commissioned by the government, many of them by federal regional governments. Finally, the third strand in his rich trajectory comprises mostly female nudes. Many of those nudes were greatly appreciated in French art salons, gaining him recognition and distinction. He passed away in 1937, leaving a great body of work and an important legacy to Brazilian art.

Brazilian Art history, critics, and audiences have acknowledged the role Parreiras played in making available those purportedly mythical natural landscapes, images of the nation, its peoples, and the role of the Brazilian nation in the larger picture of the Americas. In critic Angyone Costa's view,

> [Parreiras] paints the Brazilian landscape like nobody else did before. He paints the smooth and sometimes ecstatic landscape of the southern mountainous lands, as he will later paint, and perform the miracle, hitherto unrealized, of reproducing on canvas the wild tangle, the polychromia phantastica, the barbaric, frightening intertwining of the Amazonia.
>
> *(1927, 72)*

Parreiras' masterful and accurate evocation of the particulars that go with nature, land, and with the cycles of life, renewal, regeneration, decay, and extinction bring to the fore his commitment to Brazilian landscape: "Parreiras' art is a solitary manifesto against destruction" (Portella 2017, 118). Parreiras' oeuvre patently shows what has been discussed so far. These ideas are clearly exemplified in some of his most emblematic works.

The culmination of his landscape painting is probably a canvas called "Sertanejas" (see Figure 6.2). It is a large-format painting in which Parreiras' outdoor painting style is extremely refined and shows an in-depth knowledge of a wild natural area untouched by the human figure. The soft light of the forest tunnel, the stones covered with lichens, the leaves that form a dense

FIGURE 6.2 Antônio Parreiras (1860–1937). "Sertanejas," 1896. Oil on canvas. Rio de Janeiro, Museu Nacional de Belas Artes.

carpet on the damp soil contribute to an atmospheric sense of immersion within an enveloping forest where the usual human sense of orientation fails. This canvas seems to revisit the Rousseaunian idea of the "state of nature"[12] by suggesting that Nature is far larger and more beautiful than the fragmentary view a landscape may offer. Nature is, in its essence, what remains free from human intervention. And landscape is always already an act of interpretation, an imaginative act produced at the intersection between the artist's eye and the eye of the beholder.

Parreiras captures the subtle flight of the almost transparent butterflies hovering over the greenery of the forest mass. His rendering of Nature captivates the eye with his attention to detail. Some of his canvases show wild animals, landscapes of rural areas, as well as the sites where history takes place.

The canvas "Conquista do Amazonas" (see Figure 6.3) is considered a "historical" painting that still portrays nature as it appears in the first exchange between the Portuguese and the natives in the state of Pará. Parreiras takes from Impressionism the technique to represent nature using the effects of light and a variety of complementary colors to show the vividness, quickness, and spontaneity of the present moment. Most impressionist painters painted in plein air. This painting was completed in Niterói in 1907 and seeks to portray the first voyage of the Portuguese along the Amazon River, which lasted two years and was carried out from 1639 to 1640. The work highlights the saga of Europeans in the conquest of Amazonian lands, and it shows Parreiras' rendering the act of possession of the lands. The portrayal of Indigenous people, the sails and masts of the caravels, and the pirogues and

FIGURE 6.3 Antônio Parreiras (1860–1937). "Conquista do Amazonas," 1907. Oil on canvas, 400 × 800 cm. Belem, Museu Historico do Estado do Para.

canoes of the natives are all elements that feature alternative modes of inhabiting the world, especially regarding subsistence and social organization.

This 1907 canvas draws attention for its imposing size: it is 9 meters long and 4 meters high. "A Conquista do Amazonas" strongly contributed to reinforce the idea of a republican nation in the state of Pará. Parreiras visited Belém, the capital of Pará, at the height of the exploitation of the rubber economy. In 1905 he arrived in Belém to exhibit a series of paintings painted in Rio de Janeiro, and the governor of Pará, Augusto Montenegro, offered the main hall of the Teatro da Paz for the setting up of the exhibition. He commissioned from the painter a canvas that could represent the action of Pedro Teixeira[13] during the Amazon occupation for the Portuguese crown, demarcating the borders of the states from Spain and Portugal. Portugal would get the lands west of the Franciscana settlement founded by the Portuguese—in present-day Ecuador—and the remaining would pass on to the Spanish crown. According to Raimundo Nonato de Castro (2010, 2019, 2022), Parreiras' canvas addresses the republican elements of the construction of a comprehensive civic pedagogy for the literate and illiterate in the Brazilian Republic, allowing all ages and races from the beginning of the 20th century to feel that they were part of the new political regime.

"A Conquista do Amazonas" is a crucial visual document that properly marks the birth of the state of Pará. The canvas was delivered in 1908 and allowed observers to see the colonial history of Pará, with its vast territorial extension, as well as verify the importance of the work for the Brazilian Republic that sought to build its identity, thus breaking with the remnants of the imperial period. As Igor Gonçalves Chaves has noted, "The Portuguese conquerors occupy the leading role and the centrality of iconography, demonstrating—as in the Foundation—the privileged place of the Lusitanian European as a carrier and builder of civility" (2016, 124–125). In fact, the idea presented in Conquista is an epic feat, the first milestone in the history of Pará-Amazon and worthy of being remembered, because it is the moment in which "the adventurous genius, which so characterized the Lusitanian race of those famous times, expanded" (Chaves 2016, 125). As mentioned before, the memorable adventurous spirit and skilled knowledge of Portuguese marine exploration and trade is once again faithfully portrayed on this canvas.

Even if, due to the occasion, the historical–social excels above the rest, the luxuriant nature of the Amazonian region is foregrounded: large Amazonian trees stand out and delimit the picture, the sunlight illuminating the scene, and the river also shows through the great boats anchored on its banks. The force of the wind moves the trees, the flags, and the sails on the vessels. An adequate strength and speed of wind was fundamental for navigation at the time of exploration and conquest. Nature, as portrayed within the canvas, is mostly under the control of the West, which began to use it in their innumerable displacements toward the New World. The canvas acts as a document that allows us to understand Brazilian history, sociality, and nature.

On the left, a small patch of Amazonian forest shows a colossal tree trunk enveloped by climbing plants standing close to leafy palm trees, and on the right, a crowd attending this act of inauguration remains attentive to the explorers. In the foreground, a series of canoes makes room for the arrival of the Portuguese to Amazonian lands. Central and marginal figures[14] fill in the natural space of the clearing, whose major occupation is the voyeuristic gaze of the newly arrived men taking possession of the land. This deceivingly peaceful scene of appropriation and redistribution of the Amazonian lands represents a misleading scene of origins prior to the emergence of the tropicopolitan ethos. Social and economic activity, administration, indoctrination, and the whole colonial apparatus will gradually gain possession of material and immaterial realities. Nothing escapes the colonial injunction, "Honour your superiors!," as the Officer in Franz Kafka's "Penal Colony" (1914) inscribes in the Condemned Man's skin.

Rather than just a static setting, there is movement in Parreiras' landscape paintings where the forces of nature find unique expression in the vast immensity of the Brazilian nation. These images have since been circulating among diverse Brazilian cultures and eerily anticipate contemporary environmental attitudes toward the nation, most often imagined as an open, inexhaustible land of mythical dimensions.

Both landscape and historical painting find in Parreiras a crucial and very prolific chronicler of Brazilian realities. Parreiras expressed several times the desire to create a narrative about the country's history based on his historical paintings. It is with a statement of this type that the painter ends the writing of his autobiography *The Story of a Painter Told by Himself*:

> From Brazilian artists to the present I have executed the greatest number of historical paintings. They should have constituted a small compendium

> of the History of Brazil, as advised by my great friend Rocha Pombo when I showed him the definitive sketches of the historical paintings that I had executed, such was the rigor of the documentation that he found in them.
>
> *(Parreiras [1926] 1999, 53)*

Ever since the first years of the Academia Imperial de Belas Artes, the search for the constitution of a national art valued above all landscape painting, making it responsible, alongside historical painting, for the exaltation of the country given the identification between Brazil and its nature. Parreiras produced his work at the juncture between the mythologization of the land and a new time when nature ceased to be the expression of the nation's idyllic ideal. At this moment, we witness authors like Euclides da Cunha marveling at the strength of the Amazon rainforest, a space relegated to the margins of history, where "that sovereign and brutal nature, in full expansion of its energies, is an adversary of man" (2006, 27). Throughout Parreiras' trajectory, his unique portrayal of local and regional landscapes in "plein air" speaks to quintessential mythical aspects of Brazilian nature—vastness, incommensurability, renewal—at a time when, in his view, the noble art of painting still retained the glorious function of educating, clarifying, and easily recording itself in the soul of the people (Salgueiro 2009).

Conclusion

Through their negotiation of continental models of painting nature and their exposure to the Brazilian land, Débret and Parreiras continually viewed themselves as artists whose work was always ethically responsive to the human and the environmental and whose permanent search for an "authentic" interaction with Brazilian localities ended up producing their unique oeuvres. Their engagement with nature and culture—with an extensive knowledge of history, literature, traditions, beliefs, habits, and concerns—in an immensely diverse country like Brazil urged them to embrace the mythical dimension of its origins in a unique Edenic land to challenge the Portuguese colonial enterprise and to celebrate Brazil's promising future. A wide range of pictorial genres, techniques, styles, skills, themes, motifs, experiences, teachings, exhibitions, museums, customers, and federal and national administrations share responsibility for shaping their trajectories and framing their respective successful careers.

As a kind of artist-in-residence in Rio de Janeiro, Jean-Baptiste Débret achieved success in his artistic career during his lengthy pictorial stage in Brazil. His adaptation, closeness, and near-symbiosis with the uses and customs of his adopted land resulted in the intelligent combination of forms, themes, and motifs that characterize his work.

Débret, an accomplished painter of court portraits, popular scenes, and special events, employed by Portuguese king Dom João VI, spent 15 years in

Brazil, becoming a major chronicler of Brazilian social history. In 1827 and 1828, Débret undertook a series of journeys along the coasts and into the interior of South Brazil. Documenting the land and its peoples, his career can be understood as an endeavor to accurately comprehend Brazilianness. His is the most extensive visual archive, witnessing "the day-to-day operations of the slave systems of Brazil, or of any part of the slave Diaspora [...]" (Wood 2014, 40).

His *Voyage Pittoresque et Historique au Brésil* appeared from 1834 to 1839 and included natural history, landscapes, topography, inhabitants, and customs. His encyclopedic scope is also a living record of the emergence of a bourgeois life in the tropics and provides a crucial vantage point to critically examine power and social relations among slaves, masters, free men, social classes, and regional unbalances and differences. Probably owing to the book's high price (colored copies sold for 416 francs), the work did not sell well in his time. Débret would later be "discovered" in the 1930s, and his popularity has grown considerably up to the present moment.

Antônio Da Silva Parreiras painted Brazil throughout his life. His dedication and passion for painting makes his work into a timely document to understand the Brazilian natural and political reality. It also offers audiences a superb vantage point from which to behold a country and a society opening up to the world, in its colonial trajectory and promise for a free and unbounded Republican future.

Parreiras encourages his viewers to identify with the landscape, to connect with the land and lived environment. His artistic vision, as demonstrated by both his landscape and his historical paintings, acquires credibility by virtue of his having been both and academic and extra-academic painter and a tireless worker for a new nation. Although some of his work might risk romanticizing the nation in its myriad forms, it is important to remember that these canvases were crafted by a devoted painter whose vision was "disrupted" by his continuous travels to Europe and engagement with the circles of academic painting in the continent. His participation in academies and salons alongside the most prominent European painters crystallized in his unmistakable style. Parreiras approaches local appearances and themes of everyday life focusing on Brazilianness as seen from inside out.

A certain pictorial counter-tropical moment is what enables artists, audiences, and critics to unsettle and resist the hegemony of Western models embedded within narratives of domination, subjugation, and conquest. Both Débret and Parreiras in their respective neoclassical, post-romantic and premodernist esthetic make room for the emergence of new social and subjective (racialized) identities born out of other landscapes in an unruly Brazilian nature. Both painters struggled to remedy the impact of colonial ways of seeing that perversely distorted perceptions of land and social relations (sertão, coastal areas, Amazonia, and the white Portuguese imperial subjects in relation to blacks, indigenous peoples, women, mercenaries, administrators, slave owners, merchants, and good bourgeois, among others). On their canvases, individuals and communities inhabit alternative modernities, chronologically disrupting the Western linear way of "doing" History. Débret revisits and reimagines the Portuguese–Brazilian Empire with a critical eye for colonial domination, land and human exploitation, and its concomitant dislocated social relations. In turn, Parreiras is committed to being the chronicler of the birth of the new Republic. The deep knowledge of the Brazilian reality exhibited by both of them makes their works an irreplaceable resource for the study and reflection on modern Brazil.

The legacy of the mythical dimension of nature in tropical Brazil allows for an almost inexhaustible source of motifs and inspiration for visual representation (rainforest, sertão, backwoods, etc.) and an equally rich source of knowledge and wisdom associated to the land. The construction of a distinctive image of Brazil in the 19th-century Latin American context owes much to Débret's and Parreiras' works. Based on direct observation of empirical reality, they strongly contributed to the construction of the national imaginary, one with mythical dimensions. Both painters managed to create a place where residents, visitors, foreigners, and nationals alike could find self-recognition, identification, and shelter within the Brazilian nation. Through their use of both academic and straight nature—"en plein air"—painting techniques, which invite the viewer's empathetic identification, Débret and Parreiras offered us their potentially transforming views of Brazilian distinctiveness.

Notes

1 In Pêro Vaz de Caminha's letter, his description of the encounter with the land and is peoples describes in detail how on April 21, 1500, "Neste dia, a horas de vépera, houvemos vista de terra! Primeiramente dum grande monte, mui alto e redondo; e doutras serras mas baixas ao sul dele; ede terra chã, com grandes arvoredos: ao monte alto o capitao ôs nome—o Monte Pascoal e a terra—a Terra de Vera Cruz" (http://objdigital.bn.br/Acervo_Digital/Livros_eletronicos/carta.pdf).

2 In the context of US literature, the importance of the Edenic myth has been crucial and very influential. It has been common to many other locations in the Americas. As Krueger has argued, Some of the most interesting work to come out of American Studies has been that of the historians of the Edenic Myth; the authors who have written about "Virgin Land," the "American Adam," and the "Arcadian" and "Agrarian" myths have produced a substantial corpus of challenging scholarship. Fashioning their methods of analysis after the myth and symbol analyses of modern literary critics, they have depicted a characteristic cluster of uniquely American myths. For them, the Edenic Myth in one variant or another has formed the archetypal American cultural pattern; the faith in America as a New World Garden has been a complex symbol by which Americans have interpreted their experience and according to which they have sought to order their lives. (1973, 3) In Krueger's view, by the early 19th century, a great many Americans had come to believe in the Edenic myth.

3 For a translation into English of the "Canção," see https://allpoetry.com/Cano-do-exlio.

4 Much has been discussed on Montaigne's written sources for the writing of "Des cannibales" and his second essay on "savage populations," "Des coches" (Lutri 1975).

5 In S. Aravamudan's important theorization, he asserts, I would like to propose the term tropicopolitan as a name for the colonized subject who exists both as fictive

construct of colonial tropology and actual resident of tropical space, object of representation and agent of resistance. In many historical instances, tropicopolitans-the residents of the tropics, the bearers of its marks, and the shadow images of more visible metropolitans-challenge the developing privilege of Enlightenment cosmopolitans. / Yet tropicopolitans are also the vehicle of the metaphor that inaugurates the self-valorization of the metropolitan subject. (1999, 4)

6 After Napoleon's fall, a group of French artists arrived in Rio in 1816 as members of what became known as the French Cultural Mission. As Darlene Sadlier summarizes, "The specific assignment of the French Cultural Mission was to found an Institute of Fine Arts, where students would learn the fundamentals as well as latest trends in art and architectural design" (2008, 112). Led by historiographer Henri Lebreton, the mission included landscape painter Nicolas Antoine Taunay, his brother and sculptor Auguste Marie Taunay, architect Auguste Henri Victor Grandjean de Montigny, sculptor Marc Ferrez, engraver Charles Simon Pradier, and, foremost, "one of Napoleon's favorite portrait painters, Jean Baptiste Débret" (112).

7 The work was published in three volumes. References in the text are to Volumes 1 and 2 as Débret ([1865] 1965a) and ([1865] 1965b), respectively. All references to the text and images of *Voyage Pittoresque* are taken from the digitized facsimile version found online through Rice University. All page numbers refer to the digitized page numbers, not to the numbering of the original text. All translations of the French original are the authors'.

8 Postcolonial theories and new materialism, among other trends, argue that there exists a dangerous faith in human exceptionalism, the root of both progress (for a few) and backwardness (for the rest). The idea of "natureculture" is emerging as part of new elaborations on the place of the human in the midst of all other entities and realities within the planet. The superiority of the human is no longer taken for granted and, rather, entanglements, symbiotic collaborations, and ecosystemic perspectives are privileged. Donna Haraway has been an enormously influential researcher, a historian of science, and philosopher who has greatly contributed to this field. Her ideas on entangled multispecies histories have richly expanded this area (Haraway 2007).

9 In the paradigm mapped out by Linebaugh and Rediker in *The Many-Headed Hydra* (2000), these works were primarily carried out by members of the transatlantic proletariat (the hewers of wood and drawers of water who built the infrastructure of European cities both in the metropole and colony), which, at this point in time, were largely enslaved African peoples.

10 Other descriptions of different merchandise and vendors can be found (see, for example, Débret [1865] 1965b, 57, 122, 167).

11 As it is amply known, black slaves in the New World used musical rhythms and song to communicate collectively across languages and cultures of origin. They share these forms and channels of communication with the women Débret is portraying (Metcalfe 1970; Baker 1984).

12 Rousseau portrays the state of nature as the original situation where human life would be free of all encumbrances. In such a state, humans are distinguished from other animals by two exclusive characteristics: perfectibility, the capacity to learn that could "draw [man] from his original condition" and potentially be "the source of all of [his] miseries" (1997, 141), and free will, the ability "not to be governed solely from appetite" (Bertram 2012).

13 Pedro Teixeira was a Portuguese explorer (1587–1641), the first European to travel up the Amazon River and who claimed the entire valley for Portugal.

14 Raimundo Nonato de Castro has clearly identified "'The three figures of friars, a mercenary and two Franciscans, one of them the chronicler of the expedition, Christobal d'Acuna.' At the center of the screen, they gained privileged prominence, reinforcing the strength that the Catholic Church had at that time" (2019, 113).

References

Álvarez, José Maurício Saldanha. 2009. "'Era terra do Brasil:' Representação da nação brasileira na obra de Antônio Parreiras." *19/20* 4, no. 2 (April). http://www.dezenovevinte.net/obras/ap_jmsa.htm.

Anderson, Benedict. 2006. *Imagined Communities: Reflections on the Origin and Spread of Nationalism*. Rev. ed. London: Verso.

Aravamudan, Srinivas. 1999. *Tropicopolitans: Colonialism and Agency, 1688–1804*. Durham, NC: Duke University Press.

Baker, Houston. 1984. *Blues, Ideology and Afro-American Literature: A Vernacular Theory*. Chicago: University of Chicago Press.

Bertram, Christopher. 2020. "Jean Jacques Rousseau." In *The Stanford Encyclopedia of Philosophy (Winter 2020 Edition)*, edited by Edward N. Zalta. https://plato.stanford.edu/archives/win2020/entries/rousseau/.

Bishop, Elizabeth. 2011. *Prose*, edited by Lloyd Schwartz. New York: Farrar, Straus and Giroux.

Burke, Peter. 2001. *Eyewitnessing: The Uses of Images as Historical Evidence*. Ithaca, NY: Cornell University Press.

Caminha, Pêro Vaz de. 2019. "Letter to King Manuel I of Portugal." In *The Brazil Reader: History, Culture, Politics*, edited by James N. Green, Victoria Langland, and Lilia Moritz Schwarcz, 2nd ed., 12–17. Durham, NC: Duke University Press.

Carvalho, José Murilo de. 2001. "Nación imaginada: memoria, mitos y héroes." In *Reflexiones en torno a 500 años de historia de Brasil*, edited by Elda González, Alfredo Moreno, and Rosario Sevilla, 87–113. Madrid: Catriel.

Castro, Raimundo Nonato de. 2010. "O quadro Conquista do Amazonas de Antônio Parreiras e a idea da nação." *19/20* 5, no. 4 (Oct–Dec). http://dezenoveveinte.net/obras/ap_rnc.htm.

Castro, Raimundo Nonato de. 2019. "'A conquista do Amazonas:' Antônio Parreiras e sua representação de Amazônia." *Nova Revista Amazônica* 7, no. 2: 97–116. http://dx.doi.org/10.18542/nra.v7i2.7508.

Castro, Raimundo Nonato de. 2022. *História e Narrativa Visual na Tela A Conquista do Amazonas de Antônio Parreiras (1905–1908)*. Belém, Brazil: Instituto Federal do Pará.

Cerdera, Fábio. 2012. "Práxis enunciativa: relações interdiscursivas e criação plástica na obra de Antônio Parreiras." *19/* 20 8, no. 3. www.dezenovevinte.net/obras/ap_praxis.htm.

Chaves, Igor Gonçalves. 2016. "A Fundação da cidade de Belém inventada pelos pinceis artísticos." *Hydra* 1, no. 1: 118–130. https://doi.org/10.34024/hydra.2016.v1.9118.

Clark, Kenneth. 1949. *Landscape into Art*. Boston: Beacon Press.

Costa, Angyone. 1927. *A Inquietaçao das Abelhas*. Rio de Janeiro: Pimenta de Mello & Cia.

Cunha, Euclides da. 2006. *A margen da história*. São Paulo: Martin Claret.

Débret, Jean-Baptiste. (1865) 1965a. *Voyage au Brésil*, Vol. 1. Paris: Firmin Didot Frères. 3 vols. Facsimile edition. Rio de Janeiro: Distribuidora Record. Digitized edition. https://hdl.handle.net/1911/70167. Citations refer to the digitized edition.

Débret, Jean-Baptiste. (1865) 1965b. *Voyage au Brésil*, Vol. 2. Paris: Firmin Didot Frères. 3 vols. Facsimile edition. Rio de Janeiro: Distribuidora Record. Digitized edition. https://hdl.handle.net/1911/70172. Citations refer to the digitized edition.

Descola, Philippe, and Gísli Pálsson, eds. 1996. *Nature and Society. Anthropological Perspectives*. London: Routledge.

Eckl, Marlen. 2015. "'Everywhere Paradise Is Lost:' The Brazilian National Myth in the Works of Refugees of Nazism." In *KulturConfusão—On German-Brazilian Interculturalities*, edited by Anke Finger, Gabi Kathöfer, and Christopher Larkosh, 219–246. Berlin: De Gruyter.
Gonçalves Dias, Antônio. 1998. *Poesia e Prosa Completas. Primeiros cantos*, edited by Alexei Bueno. Rio de Janeiro: Editora Nova Aguilar.
Haraway, Donna J. 2007. *When Species Meet*. Minneapolis: University of Minnesota Press.
Hulet, Claude L. 1995. "The Columbus Letter of February 15, 1493, and the Pêro Vaz de Caminha Letter of May 1, 1500: A Comparison." *Mester* 24, no. I (Spring): 107–124.
Krueger, Thomas. 1973. "The Historians and the Edenic Myth: A Critique." *Canadian Review of American Studies* 4, no. 1 (Spring): 3–18.
Linebaugh, Peter, and Marcus Rediker. 2000. *The Many Headed Hydra: Sailors, Slaves, Commoners, and the Hidden History of the Revolutionary Atlantic*. Boston: Beacon Press.
Lutri, Joseph R. de. 1975. "Montaigne on the Noble Savage: A Shift in Perspective." *The French Review* 49, no. 2: 2016–2011.
Metcalfe, Ralph H. 1970. "The Western African Roots of Afro-American Music." *The Black Scholar* 1, no. 8: 16–25. http://www.jstor.org/stable/41206250.
Mitchell, W.J.T., ed. (1994) 2002. *Landscape and Power*. Chicago, IL: University of Chicago Press.
Montaigne, Michel de. 1993. *Essays*, translated by J.M. Cohen. London: Penguin.
Parreiras, Antônio. (1926) 1999. *História de um pintor. Contada por ele mesmo*, 3rd ed., edited by Maximiano de Carvalho e Silva. Niterói: Niterói Livros.
Portella, Isabel Sanson. 2017. "Antônio Parreiras: a floresta como ateliê—pintura direto da natureza." In *Oitocentos. Tomo IV: O Ateliê do Artista*, edited by Arthur Valle, Camila Dazzi, I. Sanson Portella and Rosangela de Jesús, 111–121. Rio de Janeiro: CEFET/RJ, DezenoveVinte.
Pratt, Mary Louise. 2002. "Modernity and Periphery: Toward a Global and Relational Analysis." In *Beyond Dichotomies: Histories, Identities, Cultures, and the Challenge of Globalization*, edited by Elisabeth Mudimbe-Boyi, 21–47. Albany: SUNY Press.
Rochette, Raoul. (1865) 1965. "Introduction." In *Voyage au Brésil*, Vol. 1, by Jean-Baptiste Débret, 1–8. Paris: Firmin Didot Frères. Facsimile edition. Rio de Janeiro: Distribuidora Record. Digitized edition. https://hdl.handle.net/1911/70167.
Rousseau, Jean Jacques. 1997. *The Discourses and Other Early Political Writings*. Cambridge: Cambridge University Press.
Sadlier, Darlene J. 2008. *Brazil Imagined: 1500 to the Present*. Austin: University of Texas Press.
Salgueiro, Valeria. 2009. "Pintor e crítico—Antônio Patrreiras n'*O Estado de São Paulo* (1894–1895)." *19/20*, 4, no. 1 (January). http://www.dezenovevinte.net/criticas/ap_vs.htm.
Schwarcz, Lilia Mortiz. 2006. "A Mestizo and Tropical Country: The Creation of the Official Image of Independent Brazil." *Revista Europea de Estudios Lationoamericanos* 80 (April): 25–42.
Wood, Marcus. 2014. "Slavery and the Romantic Sketch: Jean-Baptiste Débret's Visual Poetics of Trauma." *Journal of Historical Geography* 43: 39–48.

7

NEW COSMOGONIES OF WASTE NEGOTIATED IN THE ART OF MOHAMED LARBI RAHHALI

María Porras Sánchez and Lhoussain Simour

Introduction

Mohamed Larbi Rahhali (b. 1956, Tétouan, Morocco) is a trained visual artist, a professional technician at the Fine Arts Institute in Tétouan, and an occasional fisherman whose connection with the sea has inspired his artistic career. In many of his works, Rahhali collects and manipulates matchboxes and other recycled objects to create a range of interior and exterior views and scenes that resonate with intricate issues about life and its complexities. Through his oeuvre, he captures intimate moments of experiences and dreams that celebrate everyday Moroccan life in its secrets and intimacies and seizes in an ephemeral manner the pulsating links that unite his environment. His works have been exhibited in collective exhibitions in international museums such as Darat al Funun (Amman, 2010), La Kunsthalle (Mulhouse, 2013), the New Museum (New York, 2014), MACBA (Barcelona, 2015), Musée Mohamed VI (Rabat, 2016), and Museo Nacional Centro de Arte Reina Sofía (Madrid, 2021). His works have also been shown in several solo exhibitions in different art galleries in Morocco.[1]

This chapter attempts to provide a critical reading of Larbi Rahhali's artistic representations in their assimilationist, expressive, and subversive forms. Anchored in the context of an ongoing ambitious artistic project—*Omri* ("My Life")—that he started at the beginning of his career in 1984 and that includes an assemblage of countless matchboxes with tiny inscriptions, drawings, sculptures, and found objects, Rahhali's artworks resort to nature and set boundaries for his sensitive artistic exploration of its processes. His art captures the essence of the ephemeral and the traces of the invisible and summons the exploration of different states of waste matter. It also presents a symbolic substratum where myths are subtly embedded with a poetics of the

DOI: 10.4324/9781003348535-11

everyday. We argue that the imbrications of myth are conveyed through the notion of *baraka* ("divine blessing"), not understood as a singular myth but as the mythic fabric of Moroccan spirituality and social life.

Waste recycling campaigns in Morocco have often been conducted by both governmental and nongovernmental associations to raise public awareness of specific and complex environmental issues threatening society as a whole. In the arts, waste has become an expressive medium for contemporary debates about society and culture. Echoing environmental issues, Rahhali arranges his matchboxes to create imaginative, differently shaped installations from meticulous observation of elusive substances that not only offer so many experimental artistic and expressive bases but also become an ode to evanescent moments in life sanctioned by the secrets of nature. In Larbi Rahhali's art practice, his artistic materials, which attest to the symbolic connection that exists between nature and humankind, create minute universes of waste made of disposable objects of everyday life. *Omri*, regardless of its title ("My Life"), goes beyond the personal, evoking new cosmogonies of waste, stories of origins, and a (re-)creation of the world through the lens of the artist-fisherman, who endeavors to incorporate myriads of universes in an ongoing work. These open gateways to Morocco's past and present blur the boundaries between reality and imagination, inclusion and exclusion, personal memory and myth. This symbolic connection is enhanced by the use of myth in Rahhali's artworks as charms instilled with blessings and examples of *baraka*, a deeply rooted cultural and spiritual construct in Morocco.

Hence, if waste is evocative of uselessness, worthlessness, and defectiveness, how, in the case of Rahhali's artistic project, can it creatively be recycled to serve artistic patterns of memory, local history, resistance, and dissent? Drawing on the theories that inform the esthetics and politics of waste and on postcolonial ecocriticism as a critical strategy, this contribution seeks to understand how Rahhali ventures into a poetics of waste to evoke the fresh artistic focus on human-Earth relations today. It also seeks to explore how the cosmovisions embedded in *Omri* and other works by Rahhali, permeated by salient moments of order and chaos, identity formation, personal memory, and collective imaginary, are discursively negotiated and culturally represented in their shifting national context. Such a context instills the evocative concept of *baraka* as an abstract construct that resonates with divine blessings and magical powers in sustaining the individual spiritual beliefs. In the same vein, Walter Benjamin initiates a discussion about the distinction between auratic and non-auratic art forms. For him, an aura is "a strange tissue of space and time: the unique apparition of a distance, however near it may be," that pervades an artwork to create "a psychological inapproachability—an authority—claimed for the work on the basis of its position within a tradition" (Benjamin 2008, 14). Hence, the auratic visual texts are endorsed by their enclosed context of a traditional realm that finds expression in "a cult" or in a myth. The Moroccan traditional realm, in its complex interconnectedness with the religious and in its almost uninterpretable dynamics for Western readers, imbues Rahhali's art with intricate signifiers where the sacred and the profane collide in asymmetrical ways. *Baraka*, accordingly, becomes a mythic consolidation of the artistic aura that permeates Rahhali's artworks and bestows on his art a powerful poetics in which spirituality, myth, and everyday life converge and collide.[2]

Societal efficacy of art

In her essay *How to Do Things with Art* (2010), art historian Dorothea von Hantelmann reflects on the social and political possibilities of art and emphasizes "how artists can create and shape social relevance" (9). This notion seemingly contradicts Adorno's ideas on art's autonomy and the uselessness of giving it a social function (1997). However, von Hantelmann divests the artist of this responsibility: the societal efficacy of art is performative and only attainable within the museum. The performative in art, accordingly, "involves outlining a specific level of meaning production that basically exists in every artwork, although it is not always consciously shaped or dealt with—namely its reality-producing dimension" (2010, 18–19). Only when the artwork is exhibited might societal efficacy take place, when art enters in direct contact with the viewer, who in turn interprets the artwork.

Drawing from Judith Butler's ideas on performativity, von Hantelmann explains that the museum becomes the place in which the artist's and the spectator's subjectivities mirror and signify each other: "the exhibition (and the art exhibition in particular) actively constructs a relationship between the production of subjectivity and the production of material objects. On the one hand it exhibits objects derived from the subjectivity of the artist, and on the other it presents them in a way that seeks to have maximum impact on the subjectivity of the viewer" (2010, 11). Therefore, the social function of art is not created by the artist but by a contract between subjectivity of the artist and that of the viewer within the space of the museum.

As Pieprzak argues, much of contemporary art in Morocco is exhibited outside state-run museums. That is why in the last ten years, a new generation of curators and artists have become entrepreneurs and have created new "tactical spaces of memory and shared resources to fill the museum void" (2013, 440): small galleries, ephemeral open-air museums, artistic expeditions,[3] and art biennales. For Pieprzak, this situation has emancipatory possibilities because artists and curators have become more critical of the past and less subject to the dictates of the state-sanctioned narratives of modern art history in Morocco. That is why artists such as Larbi Rahhali, whose artworks have been exhibited in key spaces in this new ecosystem, such as L'appartement 22 in Rabat (2015), Voice Gallery in Marrakech (2018), and the now extinct Saida gallery in Tétouan (2017), are examples of a Modernist Moroccan art in which "the concept of participation is key" (Pieprzak 2013, 435).

In addition, the notion of societal efficacy of art is especially apt in Rahhali's *Omri*. Every time this artwork is exhibited in public, the artist creates new installations, including different compositions that show only a selection of the countless matchboxes that Rahhali has made over 40 years. In *Omri*, the contact between the viewer and the artwork is vital and it takes place not only in a single museum but in the different spaces in which the different versions of this work have been exhibited. *Omri*'s performative possibilities open up social efficacy by raising awareness: that is, as we will argue, an ability to evoke an environmental engagement with the viewer prompted by "the pulsating visuality" of the pictorial representation (von Hantelmann 2010, 38).

The role of artists in saving the planet might be overestimated, their real potential to do it disregarded as a well-meant overstatement. However, as Lucy Lippard suggests, artists can become allies of this change through their aesthetic and critical engagement with life:

> [Artists] can't change the world … alone. But with good allies and hard work, they can collaborate with life itself. Art can offer visual jolts and subtle nudges to conventional knowledge. At best it can lead us to look, to see, to understand, and then to act. Artists can also deconstruct the ways we're manipulated by the powers that be and help open our eyes to what we have to do to resist and survive.
>
> *(Lippard 2021, 47)*

For art historian T.J. Demos, environmentally engaged art has a double potential: "to both rethink politics and politicize art's relation to ecology, and its thoughtful consideration proves nature's inextricable binds to economics, technology, culture, and law at every turn" (2016, 10). This chapter aims to explore how the artistic practice of Mohammed Larbi Rahhali explores this double potential and expands viewers' vision by inviting us to resist and survive in a world marked by anthropocentrism and ecocide. It also alludes to *baraka* as a mythic component of Moroccan culture transmitted through social ties.

Environmental politics in Morocco and the poetics of the everyday

Morocco has been working since 2013 to develop a national environmental legal framework (United Nations Economic Commission for Europe 2021) and is one of the few countries with a nationally determined contribution (NDC) in line with the target of limiting global warming by 1.5°C. Recently, Aziz Rabbah, Morocco's ex-minister of Energy, Mines, and Environment, praised Morocco as a "model country" for green energy in front of the United Nations Environmental Assembly and described efforts at the national level to accelerate transition to a green economy by 2030 (International

Institute for Sustainable Development 2021). These attempts to prioritize renewable energy and flood protection programs, among other green topics, clash with news of Morocco joining an international pipeline project that plans to deliver Nigerian natural gas resources to Morocco through West and North Africa (Ortiz 2021), with the evident dangers it poses for the environment (Tyson 2021). In addition, the substantial legislative efforts made by Moroccan authorities in recent years have still not been implemented and lack funding in many areas.

At a social level, there is increasing environmental awareness in Morocco. According to a recent survey carried out by the United Nations, 68 percent of Moroccans accept the reality of climate change, and over half consider it is urgent to take measures to mitigate its consequences (UNDP et al. 2021). In addition, young green activism has been praised and encouraged by UN initiatives, showing that younger generations are eager to fight global warming through their projects.[4] However, environmentally social conscious consumption is not a reality yet (Tarfaoui and Zkim 2015). Even though Morocco has problems with managing and recycling waste, this is due not to a disinterest in sustainability but rather to a lack of structures and investments to manage solid waste (Dahchour and El Hajjaji 2020). According to official data, in 2015 just 6 percent of household waste was recycled and the rest was buried, and only up to 12 percent of industrial waste was recycled. Now Morocco's official target is 20 percent of recycled waste by 2022, a date recently pushed back to 2030 (France 24 2021). Nevertheless, national, local, private, and nongovernmental organization-led strategies have been deployed in the last few years to enhance the management of solid waste and promote recycling programs, "improving the collection and the management of solid waste taking into consideration the socioeconomical status of pickers" (Negm and Shareef 2020, 3).

Environmental awareness is currently increasing in Morocco, but Rahhali has been working with waste for almost 40 years. He anticipated a concern that was not part of the public agenda, creating artworks that established a dialogue with waste through the use of discarded matchboxes that he found in the medina of Tétouan and also fishnets and other objects recovered from the Mediterranean coast. By doing so, he was seeding an environmental sensitivity that is blooming at present. This is both implicit in the artist's praxis and explicit through his words: "These works are my message to the world, to draw the attention of the authorities and the keepers or the environment to the threat posed by natural phenomena when the environmental balance is not preserved" (Rahhali, email to authors, January 26, 2022).[5] Thus, the artist recovers waste, and by morphing it into art, he gives it back to the world with an environmental message: a literal and symbolic recycling of items that are unwanted and, yet, preserve bits of life stories and culture. In Rahhali's artworks, as well as in the notion of *baraka*, art, society, and human experience are linked. As we will show, this mythic fabric is fundamental for the Moroccan "world view and an ethos" (Geertz 1968, 36).

Rahhali's artifacts communicate their ordinary and arbitrary nature as waste objects; however, the human experience, which he himself incarnates, and the "truth-content" that pervades the manifestations of the matchbox as an upcycled object enhance the overt meanings of his artworks in their uniqueness and individuality. Also, by drawing attention to discarded materials, Rahhali takes part in a general trend current among North African artists who turn their gaze to their immediate surroundings and focus on the most mundane and ordinary details. According to art critic Alya Sebti, these artists share a conceptual trend that she defines as a "poetics of the everyday" (2013).

This concept is evocative of the intriguing contours of the everyday within its spatial and temporal shifts. Everydayness, accordingly, becomes a site of ordinariness that reveals banality, usualness, and mundanity but that can, nonetheless, be described, interrogated, and challenged. Representing the everyday is questioning the ample multifaceted possibilities that the ordinary can engage poetically and esthetically. Artists such as Younes Baba-Ali, Hicham Berrada, and Zineb Andress Arraki "are mining the everyday as a strategy to intertwine art and life, revealing the everyday by slightly mediating its conditions" (Sebti 2013). The task of these artists is not turning the everyday into the extraordinary but to shed light on different realities, to highlight the beauty of the ordinary, or to build up connections between unlikely events or topics. Although Sebti does not mention Mohamed Larbi Rahhali, his works, frequently made of found objects and humble materials, definitely match this poetics. Artists such as Yto Barrada, Badr El Hammami, Mustapha Akrim, and Safaa Erruas, to mention a few, can also be included within this poetics. For Sebti,

> our own vision, our own experience of the everyday—as creators or viewers—is expanded by virtue of the confrontation of our reality with someone else's, stretching our imagination but also perhaps our empathy well before we try to connect to larger events or histories.
>
> *(2013)*

Rahhali's works subtly call for an environmental awareness through the poetics of the everyday. He embellishes waste and makes the viewers interact with and interrogate the validity and origins of his materials. In so doing, Rahhali hammers esthetically and discursively on the connection between upcycling and everydayness, waste and myth, individual memory and collective history.

While addressing Rahhali's art, we also need to take into consideration the background against which his artworks are exhibited, because it also brings into focus the North–South divide. Though it is exhibited in Europe, Rahhali's work questions the role of the European and Northern Mediterranean countries in the production of detritus, often disposed of and destroyed in the countries of the Southern Mediterranean, polluting their inland and coastal areas (Issa 2021). When exhibited in Morocco and SWANA (South West

Asian/North African) countries, Rahhali's art engages with an environmental consciousness that has been growing at a steady pace in recent years. When exhibited in Europe, Rahhali's artistic project is in line with the central questions in postcolonial ecocriticism raised by Graham Huggan and Helen Tiffin:

> Is there any way of reconciling the Northern environmentalisms of the rich (always potentially vainglorious and hypocritical) with the Southern environmentalisms of the poor (often genuinely heroic and authentic)? Is there any way of narrowing the ecological gap between colonizer and colonized, each of them locked into their seemingly incommensurable worlds?
>
> *(2015, 2)*

In using the matchbox as a locally made product, Rahhali is reinforcing what Humaira Riaz calls "the aesthetics of belonging" as a major concept in postcolonial studies (2021, 75). This refers to the possibilities where home can be made through the lens of esthetics as an artistic exercise whereby oneself is explored and arranged according to a consciousness of the natural objects in the surroundings. In here, identity and belonging become relational and understood in connection with the existing waste and, further, with the context that produces and nurtures them.

The esthetics of belonging in Rahhali's artworks resonates with an emotional connection with memory that restages identity in its dynamic interactions with place. For Rahhali's generation, typical matchboxes, small-sized ones bearing the image of an Atlas lion, as a rare Moroccan breed, on the front have been familiar objects since childhood. The matchbox was used all over the country and became part of the everyday life of Moroccans before the advent of modern gas lighters. In Rahhali's project, it seamlessly merges with history, identity, and memory to highlight how he locates his relationships with nature. It has been explored "through alternative modalities of belonging, as well as ecological understandings of the relationship between human beings, the environment that surrounds them, and the other creatures with whom they share their world" (Huggan and Tiffin 2015, 135).

Sociopolitical awareness and myth in Moroccan contemporary art

After Moroccan independence in 1956, two schools of fine arts, originally founded by French and Spanish colonial authorities in Casablanca and Tétouan, paved the way for the transition toward Moroccan modernism. Modern art in Morocco was based on the conscious attempt to detach itself from the Orientalist tradition and its colonial connotations. Casablanca and Tétouan fine arts schools encouraged young artists "to depart from colonial art, to be more creative and to look for new ways to strengthen the Moroccan character of their work" (El Azhar 2016, 4). As a result, in the 1960s and

1970s, artists turned to traditional crafts for inspiration, whether using the same materials—such as brass, sheepskin, or henna, extensively used by Farid Belkahia—or motifs—as in the case of Mohamed Lamlihi, who replicated patterns found in traditional carpets in his paintings. These and other artists also started to use Arabic calligraphy and Amazigh symbols, establishing "a new visual vocabulary that broke from French colonial art education to creatively manipulate the Arabic script and the visual poetics of Sufism while integrating motifs and symbols from Morocco's Berber heritage" (Becker 2009, 143). Free from the exotifying gaze of colonialism, these modern artists could use traditional motifs and myths to shape their own vision of modernity, asserting their own cultural tradition as opposed to French-imposed models. Their approach was welcomed by a multicultural audience due to different reasons: "In their combination of the old and new, these paintings appealed, on the one hand, to the Moroccan audience, who were pleased to identify their own visual culture, and to Europeans who recognized these works as contemporary art" (El Aroussi, as cited in El Azhar 2016, 4).

Though many Moroccan artists, active in the second half of the 20th century, worked hard to create new visual models that promoted aspects of a decolonial turn in culture,[6] their art did not show the same similar emancipatory potential for confronting Moroccan authorities during the Years of Lead (1960–1980) and their aftermath. The attempts to promote a critical discourse, as those of the avant-garde magazine *Souffles* (1966–1972), were repressed. Led by Abdellatif Laâbi, Mostafa Nissabouri, Ahmed Bouanani, and Mohammed Khaïr-Eddine, the magazine started a process of cultural decolonization (Laâbi 2021). In the same period, "dissident" intellectuals such as Laâbi himself or Abdallah Zrika ended up in jail. As a result of repression, Moroccan artists working from the 1960s until the late 1990s did not criticize the official public discourse directly, showing instead signs of self-censorship (Becker 2009).

In parallel, national museums did not welcome contemporary art until late 2014, when the Mohamed VI Museum of Modern and Contemporary Art (MMVI) opened in Rabat. Even then, the museum still needed to build strong imaginaries "that would go beyond both the traditional representation of power in Morocco and the postcolonial trap of copying Western normative standards" (Dornhof 2018, 238). The historical lack of artistic infrastructure and the role of the Mohamed VI Museum of Modern and Contemporary Art within the Moroccan artistic ecosystem remain problematic for the artistic scene. This has revealed

> a sense of despair at the worsening of life conditions for the everyday Moroccan has called into question the state's investment in an "emancipatory modernity" (the potentially liberating tenets of modernity that value education, justice, and social and political freedoms).
>
> *(Pieprzak 2013, 428)*

However, Moroccan art of the last two decades has established a critical dialogue with political actions (Pieprzak 2013). According to curator and art critic Abdellah Karroum, the artistic scene of the 2000s in Morocco was "dominated by the centrality of new social and ecological issues. The globalization of communications networks brought real-time attention to the problems, both political and ecological, that confronted the people and opened up spaces for dialogue" (2021b, 305).[7] Their artistic production is characterized by the use of modest materials and unconventional media. Karroum has coined the term Generation 00 to describe this new generation of artists that show a revived interest in political issues: "Generation 00 comprises Moroccan artists living in Morocco or abroad who, in the early 2000s, began producing artworks that blended the language of art with the language of social or political issues" (2015b, 16).

Karroum has included Mohamed Larbi Rahhali's artworks in different exhibitions among those of the Generation 00, although most of these artists were born in the 1970s and 1980s. Born in 1956, the year of Independence, and educated in the School of Fine Arts in Tétouan, where he graduated in 1984, Rahhali belongs by age to the former generation of Moroccan artists. He is a bridge figure between his generation, through the substratum of traditional motifs used in his installations, but more avant-garde than his contemporaries, due to his social awareness, his "poetics of the everyday," and his use of modest materials and new media (installations) that characterize younger generations. His work is also proof of a commitment with the everyday experience based on the idea that art can transform society, a constant throughout Moroccan visual arts since independence, Karroum notes (2021a).

This is why analyzing the role of myth within his production is elusive. It is less evident than in former artists such as Farid Belkahia, Ahmed Cherkaoui, Chaïbia Talal, and Mohamed Mrabet, who used visual motifs inspired by Amazigh culture, traditional tales, tattoos, or mythic symbols such as the hand or the eye. Rather, in Rahhali's case, myth is a subtle symbolic substratum that permeates most of his artistic creations.

Cosmogonies of waste, palimpsests of memory: *Omri* and *baraka*

Rahhali's most ambitious project to date is *Omri*, an ongoing work that he started at the beginning of his career in 1984 that includes countless matchboxes with tiny inscriptions, drawings, sculptures, and found objects (see Figure 7.1). Depending on the exhibition site, Rahhali arranges his matchboxes to create installations with different shapes: they become cosmogonies of everyday life, windows to Rahhali's and Morocco's past and present. As an artist-fisherman, a collector of flotsam and discarded objects, we believe Rahhali's intention with *Omri* is no other than to create new cosmogonies of waste, giving detritus a different use to (re-)create artistic universes that tell of a story of belonging as an alternative framework for human and nonhuman relations.

FIGURE 7.1 *Omri* (1984–2021), mixed technique. Museo Nacional Centro de Arte Reina Sofía. Photograph by María Porras Sánchez. Reproduced with the artist's permission.

This artistic practice also situates Rahhali as an example of "artist as citizen," a term coined by Abdellah Karroum to describe Generation 00:

> The artist as citizen witnesses the society in which he or she lives or about which he or she is informed. The artist expresses ideas and suggests an experience of art and a way of living, proposing other possibilities, using new forms of engagement with audiences, and creating spaces around the artwork. If the idea of change is not yet there, the artwork can create expectations.
>
> *(2018, 156–157)*

The act of witnessing, engaging with other citizens, and creating new expectations out of everyday observation is central to Rahhali's practice. In the artist's words:

> I came [up] with the idea of collecting small boxes in 1982, in a café [in the medina of Tétouan], after observing how people threw their empty matchboxes away; They are very important for me, a small space where I can transport my thoughts, concepts, or whatever fragments of my culture and social tradition I see around me.
>
> *(Rahhali, email to authors, January 26, 2022)*

Rahhali's artistic principle was bearing witness to other people's waste and transforming it into an archive of impressions, cultural fragments, bits of traditions, collecting external experiences and creating a personal, and yet, collective palimpsest of memories. The artist calls his collection of matchboxes "a manuscript diary of drawings and collages with different techniques. It was a form of archiving my everyday life in an artistic way" (Rahhali, email to authors, January 26, 2022).

In that respect, *Omri* recalls Aby Warburg's *Bilderatlas Mnemosyne* (1925–1929): drawing from the memory muse, the German art historian and cultural theorist proposed a heuristic research method on memory and images. His groundbreaking project offered a compilation of simultaneous, extemporaneous images presented without hierarchization. Much like Warburg's *Bilderatlas*, Rahhali's simultaneous collection of images in *Omri* conveys the visualization of memory in an instant. It is also a sample of how contemporary Arab art offers "an unconventional form of lyrical documentary and personal reportage" producing works "in which the artist is vested with the responsibility of revising dominant historical narratives" (Gioni et al. 2014).

In addition, we find a symbolic connection with myth and Rahhali's resignified waste through *baraka*, a deep-rooted cultural and spiritual construct in Morocco. *Baraka* is not a single myth in the sense of "story," nor is it explained in detail by a specific origin story. Geertz considered that, in Moroccan culture, it was almost impossible and even a limitation "to sort out 'religion' from 'custom,' from 'folklore,' from 'myth,' from 'ceremony'" (1968, 96). Moroccan anthropologist and sociologist Abdelrhani Moundib notes that Geertz's ambiguity in defining these terms is in fact very welcome when it comes to Moroccan Islam, because such ambiguity is key to understanding religion in social and cultural history as a phenomenon in constant change, not as a "readable" semiotic artifact (as cited in Eickelman 2009, 392). In this sense, the history of "religion"—and also "myth"—is the evolving "history of the social imagination" (Eickelman 2009, 392).

For Geertz, myths can be described simply as sacred symbols (1968). "What sacred symbols do for those to whom they are sacred is to formulate an image of the world's construction and a program for human conduct that are mere reflexes of one another" (97). We share this approach to *baraka* as myth: it is an integral part of the Moroccan system of beliefs, a constant in traditional stories and everyday life occurrences that symbolizes, as Moundib highlights, "most of Morocco's cultural and historical dynamism" (2021, 66).

According to Westermarck, *baraka* is a word "used to denote a mysterious wonder-working force which is looked upon as a blessing from God, a 'blessed virtue'" (1968, i, 35). *Baraka* can be broadly translated as "blessing, in the sense of divine favor" (Geertz 1968, 44). Most myths in Morocco are related to saints and marabouts, and *baraka* appears in all of them (Geertz 1968; Eickelman 1976). It is mainly a quality attributed to saints, descendants of the Prophet, members of religious brotherhoods, and places and objects associated to them. It also works creating ties between the communities, "as a causal power transmitted through specifically contracted social ties" (Eickelman 1976, 167). Similarly, Moroccan author Abdellah Taïa reflects on the importance of *baraka* on the public perception of king Hassan II and his mythologization:

> Such a presence leaves its mark on the mind, the subconscious, and contributes greatly to myth. The myth of Hassan II focused on his baraka. Everyone in Morocco knew that he had always been incredibly lucky. He was protected, they said, by certain spirits.
>
> *(2017, 83)*

Baraka is a multidimensional concept: it might cure an illness, secure protection, or simply bring good luck. It is associated with "material prosperity, physical well-being, bodily satisfaction, completion, luck, and the aspect most stressed by Western writers anxious to force it into a pigeonhole with mana, magical power" (Geertz 1968, 44). According to Clifford Geertz, the association with magical power is not fully appropriate because it applies to this cultural phenomenon from a Western perspective and fails to describe *baraka* as a wider phenomenon exclusively characteristic of Morocco; *baraka* affects the realm of the ordinary in an extraordinary way, but its effects are circumscribed to everyday life. That is a fundamental difference between magic and myth. For Moundib, popular beliefs like *baraka* are constantly changing, as culture does, but these kinds of spiritual or cultural manifestations share a religious function since they act "as a reality relying upon how it exists in everyday life" (1999, 169). Myth is part of everyday life, of how reality is conceived and socially constructed, but magic is extraordinary and alters everyday life; it does not exist within reality but as an outside force that interrupts it. That is why identifying *baraka* exclusively with magic would be reductionist and culturally biased. Eickelman, Geertz, and Moundib seem to share this view: "the nature of religion in Morocco is the paradoxical link between *baraka*, a sense of God's grace or spirituality, and a strong secular pragmatism" (Eickelman 2009, 392).

It is important to point out that *baraka* passes to people, places, and objects in close vicinity. In a broader sense, which is especially relevant in this analysis, it can be seen as "a conception of the mode in which the divine reaches into the world" and

> More exactly, it is a mode of construing—emotionally, morally, intellectually—human experience, a cultural gloss on life. And though this is a vast and intricate problem, what this construction, this gloss, comes down to […] is the proposition […] that the sacred appears most directly in the world as an endowment—a talent and a capacity, a special ability—of particular individuals.
>
> *(Geertz 1968, 44)*

It is the transfer of human experience, this "gloss on life" in which symbolic blessings are transferred from individual to individual, in which we situate the role of Rahhali as a mediator transforming and "blessing" waste by producing recycled artworks to raise environmental awareness. Waste acquires an artistic, emotional, and political value when it is transformed by the artist. Rahhali's reused and resignified detritus evokes a precarious environmentalist blessing that interrogates the connection of waste with the dynamics of globalization and postcoloniality. The blessing of *baraka* works through a double aspect, typical of certain cultural patterns: "they are frames of perception, symbolic screens through which experience is interpreted; and they are guides for action, blueprints for conduct" (Geertz 1968, 98). Rahhali's artworks act as symbolic representations of *baraka* and also invite an ethical and ecological engagement with the spectator.

In *Omri*, which literally stands for "my life," the personal is intertwined with the collective, the collective with the environment, a common concern in eco-conscious artistic practices:

> Everything is connected, locally and globally, [...] we begin to understand the interconnective nuances between what we call nature and what we call human, which is, of course, merely another branch of nature, though we've done everything we can to separate ourselves from those parts we think we can dominate and exploit.
>
> *(Lippard 2021, 46)*

Lucy Lippard's reminder that "everything is connected" also applies to the notion of *baraka*, because it creates a virtuous chain of blessings that connect the sacred with the individual, the macrocosms with the microcosms. And, in the midst of all, the little matchboxes function as catalyzers of that transfer between the sacred and the viewer.

In addition, *baraka* has traditionally been associated with art. Moroccan decorative arts are not just utilitarian or ornamental. They have magical, protective, and sacred qualities (Jereb 2005). That is, they can be imbued with *baraka* and transfer it to the individuals using the objects.

Similarly, charms can carry *baraka* too, especially if they contain sacred words or passages from the Kur'an and they are used for preventive, curative, or positive purposes in general (Westermarck 1968). One of the powers of charms is guarding the beholder of misfortune provoked by the evil eye. Especially effective amulets are those shaped as a hand (*ḫāmsa*), an open right hand that provides a defense against the evil eye, a recurrent protection symbol in many Mediterranean countries. The number five, used in geometric patterns in charms, garments, crafts, and architecture, can also be a powerful protective device (Westermarck 1968). Eye-shaped charms are also used against the evil eye, sometimes in combination with number five or the *ḫāmsa* symbol (Westermarck 1968).

Because charms are little objects meant to be carried around for protection, often in pockets, we see a symbolic connection with Rahhali's matchboxes in *Omri*: a worthless piece of waste such as an empty matchbox, discarded after being emptied of its value, is turned into a treasured possession by the artist: "I collected them everywhere in the medina. I was absolutely thrilled by the small space in every box" (Rahhali, email to authors, January 26, 2022). The empty space becomes full of possibility, because it can be filled with a found object, a landscape, an abstract drawing, a portrait, or even a minute sculpture. For the artist, the boxes become amulets that bear witness to the stories, peoples, and traditions that inhabit the medina. By collecting them, the artist is protecting nature from waste, while at the same time preserving everyday life, archiving the memory of the city, from oblivion. The transformation of a worthless matchbox into an amulet that holds so much meaning is an artistic alchemy similar to the power of *baraka* concentrated in charms—an alchemy that reminds the spectator that life is performed and experimented in its multiplicity through art: "By animating forms of the everyday and through an insightful understanding of time that is at once deeply personal yet also extending to the planetary scale, Larbi reveals the practice of life as a constant experiment in art making" (Ginwala 2015).

If we pay close attention to each matchbox in the different installations of *Omri*, we see common motifs that repeat themselves. Because the installation at Museo Nacional Centro de Arte Reina Sofía (2021) has been the artist's largest to date—it offers a wider variety of contents—we will mostly refer to those. In his personal atlas, Rahhali seems to be trying to encompass the whole universe, from the largest element to the tiniest item. The artist uses multiple techniques: collage and drawings made with lead pencil, blue ballpoint, color felt-tip pens, and, less frequently, watercolor, a humble choice in terms of materials that visually re-creates a poetics of the everyday.

In terms of themes, we can find several drawings of the Solar System and the Earth, together with numerous landscapes (see Figure 7.2). Seascapes have a very important presence, frequently exhibiting naval vessels, some of them historical ships. Tétouan, the city where Rahhali resides, is not a coastal town, but the sea is only about 15 km away. The fascination of this artist-fisherman with the sea is clear: "I love the sea, the sea is my livelihood, I need it to survive. It is a boundless space" (Rahhali, email to authors, January 26, 2022). The sea is also evoked through the different seashell forms collected in numerous matchboxes. However, it definitely exists as a visual vocabulary that is wrapped in riddles and in enigmas, mysterious, unpredictable, and unpleasantly surprising at times.

After the vistas of the sea, the most recurrent representation is the city. In the version of *Omri* exhibited at Museo Nacional Centro de Arte Reina Sofía, the installation was crowned by a group of boxes depicting a full vista of a medina. This collage-within-the-collage mirrors the architecture of the medina itself, with its labyrinthine streets, archways, and minarets (see Figure 7.3). Other cityscapes

FIGURE 7.2 *Omri* (detail). Mixed technique. Museo Nacional Centro de Arte Reina Sofía. Photograph by María Porras Sánchez. Reproduced with the artist's permission.

present town vistas, corners, with or without human figures, sometimes drawn in detail, sometimes very schematized. Various people populate *Omri* as well: face portraits, mostly of men, coexist with stick figures in different positions. The portraits are mostly naturalistic, but monstrous figures also appear. In addition, Rahhali turns his attention to musical motifs such as musical notes and

FIGURE 7.3 *Omri* (detail). Mixed technique. Museo Nacional Centro de Arte Reina Sofía. Photograph by María Porras Sánchez. Reproduced with the artist's permission.

staves. There is also room for abstract compositions and homages to avant-garde artists, such as Fernand Léger (1881–1955), and even plain icons, such as the five-pointed star, a traditional protection symbol, as we mentioned above.

Little objects are also found within boxes. Rahhali, the fisherman who collects found objects, includes many seashells of different species. This can be seen as proof of his devotion to the sea as a source of inspiration, a homage to the

Mediterranean. But shells are also common amulets that carry *baraka* and protect the bearer from the evil eye (Westermarck 1968), reinforcing the connection between these artworks and divine powers instilled with blessings. Typically revealing is a box containing a blue eye within a shell: it reminds the viewer of a surrealistic object, of a weird pearl that returns the viewer's gaze (see Figure 7.4). But it can also be seen as a charm against the evil eye (Westermarck 1968).

However, seashells are the only organic items that we find in *Omri*. The remaining found objects are industrially made, such as a plastic comb, a plastic pacifier, plastic feathers and leaves, candies, an empty tube of glue, plastic jewelry, several coins of different currencies, an LED light, a domino, a fragment of knotted thread, a pencil, and a safety pin. In the Moroccan context, where handicrafts are such an important cultural expression, these objects, found in the medina of Tétouan or in the nearby beaches, tell a story of neocolonialism, of capitalist consumerism, of globalization.

Boxes sometimes keep their lids, so the brand is visible, showing an array of international brands compiled by the artist over decades: Moroccan (Allumaroc, Al-Namir, Le Papillon, etc.), Spanish (Vidal-Palo Largo, Tres Estrellas), British (Rothman's Tobacco), French (Nouveau Panier de Fleurs Allumettes), and even Norwegian (Nitedals Hjelpestikker). Symbolically,

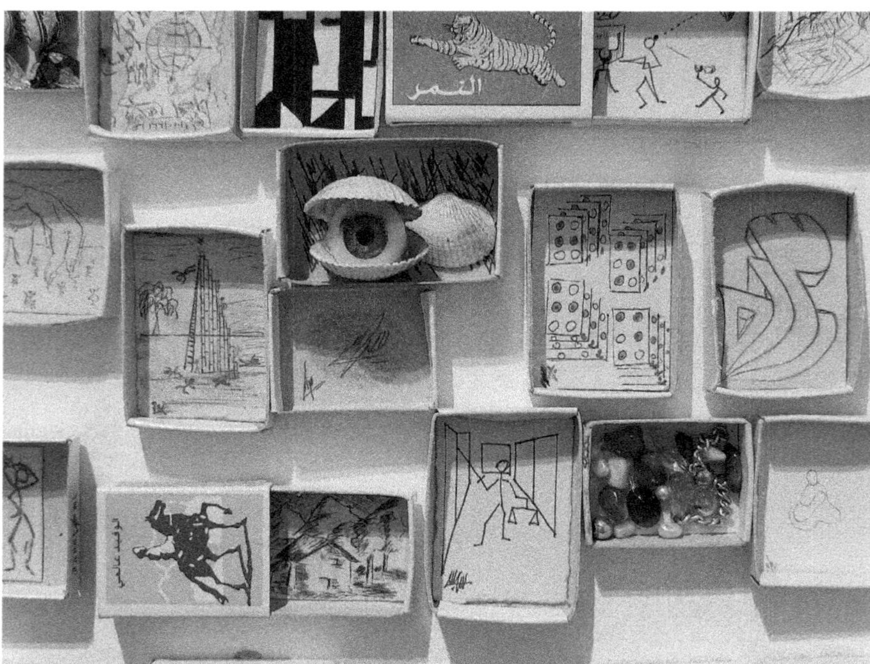

FIGURE 7.4 *Omri* (detail). Mixed technique. Museo Nacional Centro de Arte Reina Sofía. Photograph by María Porras Sánchez. Reproduced with the artist's permission.

these brands allude to Morocco as an international crossroads, but they also speak of its colonial past and evoke the contemporary neocolonialist dynamics and the imbalanced North-South relations. The arrangement of these visual cues within the overall framework interrupts global discursive strategies regarding culture and identity in a country of very sharp contrasts between modernity and tradition.

They also invoke a time that is mythical and yet contemporary. For Karroum,

> Rahhali sails the Straits of Gibraltar, the closest passage between Europe and Africa, where ancient legend has it Heracles separated the two and where modern-day legends tell of the many African children who are tempted to cross. The artist-fisherman is "witness" to the stories he illustrates in his miniatures.
>
> *(2015a)*

The act of witnessing is codified and recorded in waste objects frozen in time, divested of their previous uses and turned into tokens of this mythical and contemporary impasse.

In addition, the Warburgian character of *Omri*, where objects are not hierarchized but presented with an all-encompassing simultaneity, evoking a multiplicity of connections, allows the viewer to experience the boxes as a whole created by small units. According to the artist: "each box includes an artistic expression, but united they create a mosaic, representing a personal experience. [...] My art is my way of speaking about my culture" (Rahhali, email to authors, January 26, 2022). *Omri* is a collection of personal impressions, but as a whole, as a mosaic, it becomes an archive of a whole culture. Mosaics are an important cultural motif in Morocco and the Mediterranean area in general, so they are a form that goes beyond frontiers. The expanding nature of *Omri*, in its different installations, also re-creates the boundless space of the sea.

Omri has been exhibited at Museo Nacional Centro de Arte Reina Sofía (Madrid), Darat al Funun (Amman), MACBA (Barcelona), La Kunsthalle (Mulhouse), and the New Museum (New York), with variations in the dimensions, arrangement, and selection of boxes, albeit always installed against the museum wall. However, for one of his solo exhibitions at Voice Gallery (Marrakech), Rahhali produced several versions of *Omri* that were framed within fishnets ("Omri (série Ma Vie) #1" and "Omri (série Ma Vie) #3").[8] Through these, the fisherman-artist casts his nets and retrieves a capture of waste turned into blessed artworks imbued with *baraka*. The fishnets, as choreographic devices, also evoke the larger context that lurks behind the microcosms of the boxes.

Fishnets are materials retrieved from the Mediterranean coast. There is a clear link between the flotsam and the postcolonial legacy of a polluted Mediterranean (Issa 2021; Segalla 2021) and between such a legacy and the artist's precarious status as an artist and a fisherman. Karroum's artist-as-citizen becomes an artist-as-fisherman whose works are a mediating tool of environmental awareness, showing how postcolonial ecocriticism "performs an advocacy function both in relation to the real world(s) it inhabits and to the imaginary spaces it opens up for contemplation of how the real world might be transformed" (Huggan and Tiffin 2015, 13). In other words, the artist aims at decolonizing nature through his art:

> What would it mean to decolonize nature? [...] the cancellation of this subject–object relation between humans and the environment, the removal of the conditions of mastery and appropriation that determine the connection between the two, and the absolution of the multiple levels of violence that mediate the relation of human power over the world.
>
> *(Demos 2014, 15)*

The removal of the conditions of mastery that Demos refers to is expressed by Rahhali as a firm commitment to the environment: "Man is the first responsible of [*sic*] the earth's wellbeing and human life. These [environmentalism and commitment] are the themes I work with" he declares (Rahhali, email to authors, January 26, 2022). This commitment is also reflected in different works apart from *Omri*. In his solo exhibitions *Atrapar las impresiones* ("Catching Impressions") in Saida Art Contemporain (Tétouan, 2016), *Mind Moves Matter* in L'appartement 22 (Rabat, 2015), and *Microcosms* in Voice Gallery (Marrakech, 2018), the artist presented artworks that establish a dialogue with the cosmos, investigating its macro and micro versions. In L'appartement 22 and Voice Gallery, he presented a similar installation, titled "L'univers" ("The Universe," 2015), in which he worked with different media, including a blackboard with a chalk painting of planets and orbits and the word "RESPONSABILITY" [*sic*].[9] World maps and microscopes occupy a nearby table, evoking an examination of the Earth's problems, connecting the macrocosms with the microcosms.

Another way to achieve this connection is by producing large versions of objects associated with smallness, such as the large clothes peg that he exhibited at Voice Gallery (L'intégration #1, 2018, L'intégration #2, 2018, L'intégration #3, 1982). Planets are also represented, such as in site-specific paintings in the exhibition *Mind Moves Matter*, and in sculptures and installations, such as the ones exhibited in *Atrapar las impresiones*. In sculptures such as "New Horizon" (c. 2016), Rahhali presents a globe that is turned 90 degrees, contemplating the possibility of an Earth that changes the rotation angle, displacing the equator.[10] In the installation "Chance Planet" (c. 2016) he re-creates a bingo in which a new planet is drawn at random. These works show both environmental awareness and the need to replace ethnocentrism for more inclusive cosmovisions. As Ginwala points out, Rahhali's works are conceived to evoke "human-earth relations today [...] choreographic devices between order and chaos, the sea horizon and interstellar space, personal memory and a collective psyche" (2015). Larbi Rahhali also uses orbs in other works, such as the installations presented in L'appartement 22 for the exhibition *Mind Moves Matter*; for instance, a kinetic sculpture including a recycled fishnet, a toy orb, and several toy planes and helicopters.

Fishnets are indeed recurrent in the works of Rahhali; apart from framing some versions of the polymorphous *Omri*, they appear as a sort of canvas in other installations; that is, they are the base that articulates the rest of the artwork. This is the case of "El arco" (c. 2016, exhibited in *Atrapar las impresiones*), several untitled installations at the L'appartement 22 exhibition, and "Sans titre" (2018) and "Création" (2018), exhibited at the Voice Gallery. These nets are the perfect metaphor of the workings of the fisherman-artist: the artist-citizen that captures objects, impressions, and waste and fashions works of art from them.

Time, "perhaps the most crucial element in our understanding of climate disruption in this age of instant gratification" (Lippard 2021, 46), is also an important theme in Rahhali's work. The slow pace employed to create *Omri* (1984–ongoing) is related to a sustainability based on recycling. But the artist also establishes a direct dialogue with time in some of his sculptures, in which he creates handmade sundials made of stone, plaster, or clay ("Cadran solaire #1," 2015; "Cadran solaire #2," 2015; and "Cadran solaire #3," 2015, all exhibited at the Voice Gallery). These sundials call forth a slow, ecocentric time in which the Earth, the Sun, and the cosmos establish a balanced relation.

Conclusions

This chapter has attempted to reflect on the complex ramifications of Rahhali's artistic project. In rescuing the matchbox from the palimpsest of waste, the Moroccan artist is engaged in a process of reinscribing his life details with astute artistic shrewdness and perceptive (I)eye through the alteration of isolated and disregarded objects that have long shaped his individual and collective consciousness.

Exploring a poetics of the everyday that characterizes the younger artists of the Generation 00, Rahhali is the link between the social awareness of younger generations and the interest in traditional motifs of the earlier avant-garde artists, offering a vantage point of contemporary art in Morocco and the societal efficacy of art in general.

In addition, Rahhali's shapeshifting artworks escape any anthropocentric and ethnocentric dynamics: the minute nature of matchboxes and found objects in *Omri* reduces human scale to episodic pieces of waste. By doing so, the artist moves toward an ecocentric cosmovision. His artworks become amulets, memory keepsakes that conform an atlas of simultaneous images and contexts, charms that assert the value of waste for reimagining the past, acknowledging the present and imagining a different future in a constant blessing that links the macrocosm and the microcosms.

There is an emerging agency in the artistic practice of Mohamed Larbi Rahhali. His works are not exotic souvenirs, remnants of an Orientalist construction of Morocco, but meaningful keepsakes of things that otherwise would pass unnoticed or would be lost forever. The artist as citizen and fisherman develops a poetics of the everyday that is humble and yet all-encompassing, an environmentally aware cosmovision in which everything—waste, peoples, earth, universe, sea, objects, city—is reduced to amulets, ordinary but also precious, connecting the planet with its dwellers through a continuous *baraka*.

The *baraka* that permeates Rahhali's work is elusive and creates a knot of ambivalent meaning as to the mystic and mythical connotations it purports. In Moroccan collective imagination, even the useless elements of nature have their own *baraka*. Collecting disposable cans and plastic bottles and selling them to make a living is a blessing; their *baraka* stems from the profit they yield, which helps to sustain a family. In the case of Rahhali, the matchbox is also a *baraka* to rise to fame and get recognized as an artist. Fishing, as an everyday practice, is also endowed with a *baraka* as a way to make a living. A fisherman is not certain to land enough fish to sell at the end of the day but, as is commonly believed, each day has its own *baraka*.

The matchbox and fishing, in Rahhali's artistic vision, resonate with the divine ability to emit and glow with a *baraka*, which is almost captured in a definition as "personal presence, force of character, [and] moral vividness" (Geertz 1968, 44). The myth of the *baraka,* in the sense that it is a durable belief constructed over divine and mystic acceptances, bestows Rahhali's work with a spiritual dimension. Waste becomes a symbol of endless spiritual incarnations with powerful connections to natural objects in the surroundings.

In gaining some esthetic value and artistic merit, matchboxes in Rahhali's project cease to be trash as such; rather, they are a compilation of motifs and signifiers where life and memory, waste and art, history and society converge in a meaningful manner. The reconfiguration and relocation of waste into art acquires an intrinsic relation of ambivalent discourses about avant-garde artistic expression in Morocco. Life details in Rahhali's matchboxes become suggestive of resurrection and awareness. The waste turns into lively objects and saves the history of the matchbox in Morocco from oblivion. Through the matchbox, a waste object, medinas, minarets, streets, and city gates are resurrected; these intensify the symbolic markers of the artist's cultural identity. Rahhali's project offers a new conceptualization of the matchbox, while associating it with other categories such as space and time, which adds a web of complexities to his artistic vision and practice.

Notes

1 *Mind Moves Matter* in L'appartement 22, curated by Natasha Ginwala (Rabat, 2015), *Atrapar las Impresiones/Atrapper les impressions* in Saida Art Contemporain, curated by Carlos Jiménez (Tétouan, 2017), and *Microcosms* in Voice Gallery (Marrakech, 2018).

2 We understand Moroccan spirituality as the set of beliefs that conform the traditional, unorthodox Islam in Morocco that are "interwoven in the very structure of people's consciousness and daily interaction" (Moundib 1999, 174). *Baraka*, like the cult of saints and the presence of religious brotherhoods, most of which are bearers of and bestowed with divine blessings, is an important element of Moroccan spirituality, not as a mythic element belonging to orthodox mainstream Islam but as a powerful signifier that feeds spirituality and sustains its workings.
3 Led by Abdellah Karroum, *Expèditions du bout du monde* (End-of-the-World Expeditions), was an artistic project that consisted of a series of expeditions to the Rif mountains and other natural areas where the artists produced their work, then exhibited them in venues such as L'appartement 22 in Rabat. See, for instance, *Expèdition du bout du monde #1*, with Younès Rahmoun and Jean-Paul Thibeau, Champs de l'Arabe (Rif), 2000. http://appartement22.com/spip.php?article261. See also Karroum, *Sous nos yeux* (2015b, 16–22).
4 https://news.un.org/en/story/2022/01/1110302.
5 Authors' translation (Mohamed Larbi Rahhali's email to the authors was originally written in Spanish). We provide our translation of the subsequent references to this email.
6 In his work *Maghreb Pluriel*, Moroccan critic Abdelkebir Khatibi argues for "une pensée plurielle qui ne réduise pas les autres (sociétés et individus) à la sphère de son autosuffisance" [a plural thought that does not reduce others (societies and individuals) to the sphere of self-sufficiency] (1983, 18). Concomitant with the "pensée plurielle" that would serve for the creation of a new cultural dialogue based on difference, Khatibi evokes decoloniality to question and problematize the inherent histories of subjugation emerging from the colonial period. Such subjugation felt from within is perpetuated by the ruling bourgeois class, which reproduces colonial attitudes and politics of division and replicates colonial strategies of power and domination.
7 Translation from: https://universes.art/en/specials/moroccan-trilogy-1950-2020.
8 See http://www.voicegallery.net/photogallery/category/38-microcosms.html.
9 http://appartement22.com/spip.php?article403&id_document=1267#documents_portfolio.
10 All images of the exhibition *Atrapar las impresiones* are available at the catalogue: https://issuu.com/saidaartcontemporain/docs/catalogo_med_larbi_rahhali.

References

Adorno, Theodor W. 1997. *Aesthetic Theory*. London: Bloomsbury.

Becker, Cynthia J. 2009. "Art, Self-Censorship, and Public Discourse: Contemporary Moroccan Artists at the Crossroads." *Contemporary Islam* 3: 143–166. doi:10.1007/s11562-009-0085-z.

Benjamin, Walter. 2008. *The Work of Art in the Age of Its Technological Reproducibility, and Other Writings on Media*, edited by Michael W. Jennings, Brigid Doherty, and Thomas Y. Levin, translated by Edmund Jephcott, Rodney Livingstone, Howard Eiland, et al. London: Belknap Press.

Dahchour, Abdelmalek, and Souad El Hajjaji. 2020. "Management of Solid Waste in Morocco." In *Waste Management in MENA Regions*, edited by Abdelazim M. Negm and Noama Shareef, 13–34. Berlin: Springer.

Demos, T.J. 2014. "Decolonizing Nature: Making the World Matter." In *World of Matter*, edited by Inke Arns, 14–25. Berlin: Sternberg Press and Hartware MedienKunstVerein.

Demos, T.J. 2016. *Decolonizing Nature: Contemporary Art and the Politics of Ecology*. Berlin: Sternberg Press.

Dornhof, Sarah. 2018. "Exhibiting Contemporary Moroccan Art: Situated Curatorial Narratives and Institutional Frames of Globalization." In *Situating Global Art*, edited by Sarah Dornhof, Nanne Buurman, Birgit Hopfener, and Barbara Lutz, 231–254. Bielefeld: transcript Verlag.

Eickelman, Dale F. 1976. *Moroccan Islam: Tradition and Society in a Pilgrimage Center*. Austin: University of Texas Press.

Eickelman, Dale F. 2009. "Not Lost in Translation: The Influence of Clifford Geertz's Work and Life on Anthropology in Morocco." *The Journal of North African Studies* 14, nos. 3–4: 385–395. doi:10.1080/13629380902924034.

El Azhar, Samir. 2016. "Visual Arts in the Kingdom of Morocco." *THIRD TEXT. Critical Perspectives on Contemporary Art and Culture* (October): 1–6. Accessed January 15, 2022. http://thirdtext.org/visual-arts-morocco-bookreview.

France 24. 2021. "Firm Transforms Waste as Morocco Faces Trash 'Time Bomb.'" France 24, December 19, 2021. Accessed January 15, 2022. https://www.france24.com/en/live-news/20211219-firm-transforms-waste-as-morocco-faces-trash-time-bomb.

Geertz, Clifford. 1968. *Islam Observed: Religious Development in Morocco and Indonesia*. Chicago: University of Chicago Press.

Ginwala, Natasha. 2015. *Mohamed Larbi Rahhali: Mind Moves Matter* (exhibition portfolio). Rabat: L'appartement 22. Accessed January 15, 2022. http://appartement22.com/index/spip.php?article403&id_document=1276#documents_portfolio.

Gioni, Massimiliano, Negar Azimi, and Kaelen Wilson-Goldie. 2014. "The New Museum Presents 'Here and Elsewhere,' a Major Exhibition of Contemporary Art from and about the Arab World." New Museum. Accessed January 15, 2022. https://www.newmuseum.org/exhibitions/view/here-and-elsewhere.

Huggan, Graham, and Helen Tiffin. 2015. *Postcolonial Ecocriticism: Literature, Animals, Environment*. London and New York: Routledge.

International Institute for Sustainable Development (IISD). 2021. *Earth Negotiations Bulletin: A Reporting Service for Environment and Development Negotiations* 16, no. 156. February 26, 2021. Accessed January 15, 2022. https://enb.iisd.org/sites/default/files/2021-02/enb16156e_0.pdf.

Issa, Sintia. 2021. "Waste You Can't Deny: A Slow Trans-aesthetic in the Blue Barrel Grove." In *The Routledge Companion to Contemporary Art, Visual Culture, and Climate Change*, edited by T.J. Demos, Emily Eliza Scott, and Subhankar Banerjee, 108–118. London and New York: Routledge.

Jereb, James F. 2005. *Arts and Crafts of Morocco*. London: Thames and Hudson.

Karroum, Abdellah. 2015a. *Sentences on the Banks and Other Activities. Journal 01*. Amman: Darat al Funun-The Khalid Shoman Foundation. Accessed January 15, 2022. https://daratalfunun.org/wp-content/uploads/2015/11/Sentences_EN.pdf.

Karroum, Abdellah. 2015b. *Sous nos yeux (Ante nuestros ojos/Before Our Eyes)* (exhibition catalogue). Mulhouse and Barcelona: La Kunsthalle and Museu d'Art Contemporani de Barcelona (MACBA).

Karroum, Abdellah. 2018. "Generation 00: The Artist as Citizen." In *Situating Global Art*, edited by Sarah Dornhof, Nanne Buurman, Birgit Hopfener, and Barbara Lutz, 155–166. Bielefeld: transcript Verlag.

Karroum, Abdellah. 2021a. "La Generación 00, antes y después de la Primavera Árabe." In *Trilogía marroquí: 1950–2020* (exhibition catalogue), edited by Omar Berrada, Manuel Borja-Villel, Ali Essafi, Abdellah Karroum, Driss Ksikes, Abdellatif Laâbi, Abdellah Taïa, and Selma Zerhouni, 18–32. Madrid: Museo Nacional Centro de Arte Reina Sofía.

Karroum, Abdellah. 2021b. "Trilogía marroquí." In *Trilogía marroquí: 1950–2020* (exhibition catalogue), edited by Omar Berrada, Manuel Borja-Villel, Ali Essafi, Abdellah Karroum, Driss Ksikes, Abdellatif Laâbi, Abdellah Taïa, and Selma Zerhouni, 305–315. Madrid: Museo Nacional Centro de Arte Reina Sofía.

Khatibi, Abdelkebir. 1983. *Maghreb pluriel.* Paris: Denoël.

Laâbi, Abdellatif. 2021. "Periferia y revolución cultural (El ejemplo de Marruecos)." In *Trilogía marroquí: 1950–2020* (exhibition catalogue), edited by Omar Berrada, Manuel Borja-Villel, Ali Essafi, Abdellah Karroum, Driss Ksikes, Abdellatif Laâbi, Abdellah Taïa, and Selma Zerhouni, 73–80. Madrid: Museo Nacional Centro de Arte Reina Sofía.

Lippard, Lucy R. 2021. "Describing the Indescribable: Art and the Climate Crisis." In *The Routledge Companion to Contemporary Art, Visual Culture, and Climate Change*, edited by T.J. Demos, Emily Eliza Scott, and Subhankar Banerjee, 45–53. London and New York: Routledge.

Moundib, Abdelrhani. 1999. "How to Study of Religion in the Muslim World: Reflection on a Field Work Experience in the Middle of Morocco." *Scripta Instituti Donneriani Aboensis* 17, no. 2: 169–176. https://doi.org/10.30674/scripta.67270.

Moundib, Abdelrhani. 2021. "Interpretive Anthropology and Islam in Morocco: A Comparison between Geertz and Eickelman." In *Knowledge, Authority and Change in Islamic Societies*, edited by Allen James Fromherz and Nadav Samin, 61–75. Leiden: Brill.

Negm, Abdelazim M., and Noama Shareef. 2020. "Introduction to the 'Waste Management in MENA Regions.'" In *Waste Management in MENA Regions*, edited by Abdelazim Negm and Noama Shareef, 1–12. Berlin: Springer.

Ortiz, Jorge. 2021. "Morocco and Nigeria Are Getting Closer to Creating Their Own Gas Pipeline." Atalayar: Between Two Shores, December 22, 2021. Accessed January 15, 2022. https://atalayar.com/en/content/morocco-and-nigeria-are-getting-closer-creating-their-own-gas-pipeline.

Pieprzak, Katarzyna. 2013. "Moroccan Art Museums and Memories of Modernity." In *A Companion to Modern African Art*, edited by Dana Arnold, 426–444. Hoboken, NJ: John Wiley & Sons.

Riaz, Humaira. 2021. "'Aesthetics of Belonging': Construction of a Postcolonial Landscape in Daud Kamal's Poetry." In *Environmental Postcolonialism: A Literary Response*, edited by Shubhanku Kochar and M. Anjum Khan, 75–88. London and New York: Lexington Books.

Sebti, Alya. 2013. "Poetics of the Everyday." ArteEast.org. Accessed January 15, 2022. http://arteeast.org/quarterly/poetics-of-the-everyday/.

Segalla, Spencer. 2021. *Empire and Catastrophe: Decolonization and Environmental Disaster in North Africa and Mediterranean France since 1954.* Lincoln: University of Nebraska Press.

Taïa, Abdallah. 2017. "The King Is Dead." In *Another Morocco: Selected Stories*, translated by Rachael Small, 80–84. South Pasadena, CA: Semiotext(e).

Tarfaoui, Dalal, and Salah Zkim. 2015. "The Environmentally Conscious Consumption in Morocco: Myth or Reality." *Academic Journal of Interdisciplinary Studies* 4, no. 3 (December): 34–39.

Tyson, Luke. 2021. "Morocco's Environmental Sustainability on Display at United Nation." Morocco World News. March 6, 2021. Accessed January 15, 2022. https://www.moroccoworldnews.com/2021/03/336697/moroccos-environmental-sustainability-on-display-at-united-nations.

United Nations Development Programme, Cassie Flynn and Eri Yamasumi. 2021. *People's Climate Vote.* Oxford: UNDP and University of Oxford. Accessed January 15, 2022. https://www.undp.org/publications/peoples-climate-vote.

United Nations Economic Commission for Europe. 2021. *Environmental Performance Reviews Morocco: Second Review.* https://unece.org/sites/default/files/2021-10/ECE.CEP_.191.En_.pdf.

von Hantelmann, Dorothea. 2010. *How to Do Things with Art.* Zurich: JRP | Ringier.

Westermarck, Edward. 1968. *Ritual and Belief in Morocco.* Vol. I. New York: University Books.

8

DEATH IS LIFE IS DEATH IS LIFE

Continual regeneration in myth and the art of Maki Ohkojima

Keijiro Suga

Introduction: toward a forest of life

Japanese mythology is nature-based. Born of animism and the cult of nature, Japanese mythology has a pantheon of different layers of small, locally enshrined gods that ultimately were gathered around the lineage-establishing stories of the power of the emperor. Still, something remained outside this official version of mythology. Living near nature, people spontaneously reinvented mini-stories over and over again, and these retained a keen sense for natural processes.

Nature sustains itself by cyclical processes. The cycle of life means its end is always another beginning. Life culminates in death, death generates life. Such a principle is abundantly seen in Japanese mythology. Traditionally, we say in Japanese that there are eight million gods. The number only means "a lot," but this abundance is felt everywhere if you take a closer look at the land throughout Japan.

This chapter begins by outlining Japanese mythology. I will show a list of how different deities dominate different topoi. Then, among these, the myth of Ogetsuhime emerges. Her death is placed at the origin of foods. From there, I will look at the circular perception of death and life in the Japanese imagination. These are all preliminary discussions.

The main subject of this chapter is the work of contemporary artist Maki Ohkojima, whose guiding principles of creation seem to be, in my phrasing, "death gives life" and "we are multiple." Between death and life there is always a singular and crucially important moment of giving/receiving birth. Ohkojima's vision of life encourages us to see the world in a different way and opens up a broad horizon in parallel with that of the current multispecies approach in anthropology (see, for example, Yamaguchi 2014; Okuno 2022).

DOI: 10.4324/9781003348535-12

She tries to represent life as a forest of infinite relations, materially and perceptually. I think her work is at the leading edge of ecocritical art today, which is none other than an effective element of environmental activism (Okuno et al. 2021).

Outlining Japanese mythology

Myth is repetition. It is repeatedly told. Its content is repeatedly recreated. Its content (or story or set of mythemes) migrates from one epoch to another, from one place to another. With variations, its basic components remain the same and repeatedly act upon different audiences. The audience, or readers, may or may not receive the same lesson. The broad story lines could be interpreted differently, yet in this freedom of interpretation lies the power of story.

We may be able to revive the myth to shed new light on the world according to the condition we live in and the problems we face. When we are told a myth, we remain thoroughly passive. The more compelling the myth is, the more passive and inactive we are. And then, when the storytelling is over, we begin to live our active phase with a heightened consciousness of the environment. We literally see and feel the world differently. It is an enactment triggered by the myth. It also is an active interpretation of myth in which the myth becomes a verbal tool for our collective revision of the life-world.

Japanese mythology must have had a long history of oral transmission. After the introduction of writing, the broad outlines of Japanese mythology were for the most part compiled in two classics: *Kojiki* (712) and *Nihon Shoki* (720?). Other bits of ancient myth may appear in various *Fudoki* (Writings on Wind and Soil, roughly meaning Cosmography of the Land) from different regions of the Japanese archipelago. Both *Kojiki* (written in classical Japanese) and *Nihon Shoki* (written in classical Chinese) were edited for the emperors to establish the authenticity of their lineage and power. The stories revolve around the incidents in Takamagahara (Heavenly Field), where gods resided, beginning with its cosmogony and the birth of islands, and then shift to the deeds of the earliest emperors (who are, curiously enough, considered to be the direct descendants of gods).

Imaginably, these myths constitute an accumulation and amalgam with many different layers, which only look consistent because they are heavily edited with a centripetal force around the emperor. They may have incorporated earlier, more local little gods, for instance, that are enshrined locally without a wider currency. The emperors mentioned are the founders of the Yamato Court, but in ancient times there were other local kingships such as the Izumo in western Japan and others. There were, of course, smaller but powerful groups of other linguistic–cultural strains. With the expansion of the Yamato's territory, such smaller kingships were transformed into "local gods" or "un-obeying gods" and incorporated into the mythology of the Yamato kingdom.

My focus here is a myth of continuity and circularity of death and life. So, I will not go into the details of the whole gamut of Japanese mythology but will attempt a minimal retelling of the episode of Oogetsuhime. She is but a minor character in the story line of the epic of the trickster god Susanoo, yet her role is crucial because she is at the origin of foods on this archipelago. Even so, some preliminary remarks leading to her episode are mandatory.

What follows is just a retelling of the outline (see also Fukunaga 2003; Nakamura 2009). Japanese mythology has a strange beginning of, so to speak, autopoiesis of the universe. It begins with an obscure cosmogony. In the beginning there was a chaos without distinction and there was no heaven or earth. There appeared, without any external intervention, three gods: Amenominakanushi no kami, who places himself at the center of heaven and unites the universe, and Takamimusubi no kami and Kamimusubi no kami, both of them overseeing the self-generation of the universe ("no kami" is a general term meaning "god"). At the beginning, the earth was shapeless. It was only floating like a jellyfish drifting in the water. But then there appeared something sprouting and growing strongly toward the sky just like reeds in the springtime. This gave birth to two gods: Umashiashikabihikoji no kami, who embodied the upward growing force of the reeds, and Amenokototachi no kami, who represents the endless heaven-space itself. These five initial gods are heavenly gods and different from the gods of the earth.

Then seven gods of the earth appeared. From the chaos that resembled floating bits of half-melted fat on the water first emerged Kuninokototachi, god of the land to be materialized (hereafter I will eliminate the "no kami" part of the names), and then Toyokumo, who represents the process of the fat-like material solidifying and making up a wide and serene lake. These two gods were solitary (without companions), and their presence did not really show (become visible). Then comes a group of gendered gods, composed of five couples. Each couple is treated as one instance of the generative process and treated as one god. Uhijini (male) and Suhijini (female) signify the floating substance's separation into brine and soil, eventually composing a lake with sands and mud. The couple of Tsunoguhi (male) and Ikuguhi (female) shows that the mud of the lake gradually dries and spring plants shoot out and grow from the ground. The couple Ohotonoji (male) and Ohtonobe (female) shows the solidification of the wide open, firm ground. Then the male god Omodaru represents the completion of the surface of the earth without any leftover areas; his partner female god Ayakashikone is the consecration of the cry of joy "Aya ni kashi koshi" (Obviously superb and precious). After all these comes the most important and famous couple: Izanagi (male) and Izanami (female). They are the most important because they created the islands as well as the earliest protagonists of Japanese mythology. Their ritual of sexual intercourse and attempt at procreation are told in detail and quite interesting but not very relevant for us. What is relevant is the destiny of Izanagi's youngest child Susanoo and a little episode in his peregrination.

Eight million gods and their ecological geography

Landscape can be theophanic. The couple Izanagi and Izanami gave birth to a series of islands of Japan (14 in total) and a group of gods (35 of them). It is fascinating to note that these gods embody characteristics that correspond in many ways to the ecological conditions of the archipelago. Here is a list of gods the couple generated and the nature of these gods. The list may look tedious, but it needs to be carefully looked at to grasp its meaning. The nomenclature of these gods reflects, better than anything else, the ancient people's collective perception of the land. I am following the description in *Kojiki* (there is a slight difference between *Kojiki* and *Nihon Shoki*).

Ohokotooshio. Celebration of the completion of giving birth to islands.
Iwatsuchibiko. Celebration of the rocks and soil to build walls.
Iwasubime. Celebration of stones and sands.
Ohtobiwake. Celebration of gates and entrances.
Amenofukio. Celebration of the roofing work (with grass).
Ohyabiko. Celebration of roofs.
Kazaketsuwakenooshio. Protection against strong winds.
Ohwatatsumi. The governor of the ocean and sea.
Hayaakizuhiko. The governor of the estuary and rapid stream.
Hayaakizuhime. The female form of the same.

Hayaakizuhiko and Hayaakizuhime, being deities of the estuary, share the control of both river and sea. Together they generated the following water-related gods:

Awanagi. Showing that the troubled water is now calmed.
Awanami. Representing the troubled, turbulent sea.
Tsuranagi. Representing the bubbling surface of water.
Tsuranami. Almost the same as the above.
Amenomimakuri. The governor of the distribution of water.
Kuninomimakuri. The governor of irrigation.
Amenokuhizamochi. Holding the ladle to scoop water of the sky.
Kuninokuhizamochi. The same but functioning on the earth.

The work of Izanagi and Izanami to arrange the surface of the earth continues with the following gods:

Shinatsuhiko. The god of the wind, whose name means long breath.
Kukunochi. The god of trees, whose name represents the beautified form of the trunk.
Ohyamatsumi. The god of mountains.

Kayanohime. The god of the wild fields, whose name comes from thatch and pampas grass used for roofing. Also known under the name of Nozuchi.

The last two of the above, Ohyamatsumi and Nozuchi, are governing gods of mountains and wild fields, so they shared responsibilities of the wilderness and generated the following gods:

Amenosazuchi. The governor of steep slopes, concerning the sky.
Kuninosazuchi. The same as above, concerning the earth.
Amenosagiri. The god of the boundary of a pass, concerning the sky.
Kuninosagiri. The same as the above, concerning the earth.
Amenokurado. The governor of the valley where no sunshine reaches.
Kuninokurado. The same as the above, concerning the earth.
Ohtomatohiko. The governor of the gentle slope.
Ohtomatohime. The female form of the above.

All of the above concerns geographical characteristics. Then come other gods related to other *topoi* and elements:

Torinoiwakusubune. In celebration of a boat made with *kusu* wood and that sails as swiftly as a water bird; this is the god of traffic and also called Amenotorihune (the bird boat of heaven).
Oogetsuhime. The governor of foods. (She will be discussed in detail later.)
Hinoyagihayao. The governor of fire. Also called Hinokagabiko or Hinokagutsuchi. This is a god of burning fire and when he was born his mother Izanami was burned badly in her genitalia and became bedridden.
Kanayamabiko. When Izanami was ill because of her burn, from her vomit was born this god who governs mining.
Kanayamabime. The female form of the above.
Haniyasubiko. Born from Izanami's excrement and governing fertilizer.
Haniyasubime. The female form of the above.
Mitsuhanome. Born from Izanami's urine and governing water for agriculture.
Wakumusubi. Overseeing the growth of crops.
Toyoukebime. The child of the above and the governor of many different foods.

Eventually, Izanami died from the burn she sustained when giving birth to the fire god Hinokagutsuchi. This will lead to one of the most famous episodes of Japanese mythology—that of Izanagi's descent into Yomotsukuni (the Land of the Dead) and a failed attempt at retrieving his dead wife Izanami.

What interests us is the procreation of gods by Izanagi alone (not even parthenogenesis) after the death of his wife Izanami. As before, all of the gods generated are closely associated with one aspect or the other of the management of the land, its flora, and the necessary tools of intervention in nature. The origins of places and things are explained basically as the recycling of the dead body parts of gods. It is worth noting how body parts or fluids of a dead god give birth to new gods. Sacrificial structure is implied; a god is killed and distributes its power to other newborn agents.

Izanagi kills his own child, the fire god Kagutsuchi, who caused the burn to Izanami and thus killed its own mother. Izanami severs the child's head with his long, powerful sword and the blood gushes out to give birth to a series of gods. The blood that flowed on the stones generated three gods: Iwasaku, Nesaku, and Iwatsutsunoo. These three gods celebrate the hammer for forging swords. Then the blood on the handguard of Izanagi's sword likewise dropped on the stones to give rise to three new gods: Mikahayabi, Hihayabi, and Takemikazuchinoo (the last also known as Takefutsu or Toyofutsu). These three gods celebrate the work of fire in the process of forging swords. Finally, the blood on the hilt of the sword, dropping on the ground from between the fingers of Izanagi, gave birth to two gods: Kuraokami (the dragon god to govern rain) and Kuramitsuha (the water god to govern the fresh stream of the valley).

These eight gods above were born from blood, but Kagutsuchi's body parts were used even more productively. From his severed head Masakayamatsumi was born, celebrating the rugged mountain peaks. From his chest Odoyamatsumi was born, celebrating the mountain's foothills. From his belly Okuyamatsumi was born, celebrating the deep mountain forests. From his genitalia Kurayamatsumi was born, celebrating the dark valley. From his left hand Shigiyamatsumi was born, celebrating grand full-grown trees. From his right hand Hayamatsumi was born, celebrating lower hills. From his left leg Harayamatsumi was born, celebrating mountains with mild peaks. From his right leg Toyamatsumi was born, celebrating the mountain at the end of a range.

As we can see, gods are geography, and geography is mythologized from the beginning. Gods in early Japan, taken collectively, are a nature–technology continuum. This continuity and circularity were at the basis of the collective imagination at work in this archipelago for the length of its history.

The meaning of Oogetsuhime

Oogetsuhime is but a minor character in this *roman fleuve* of mythical stories. She appears as a victim of violence exercised by Susanoo. It is important to grasp the meaning of this divine violence, and for that first we need to know who Susanoo is. Susanoo is Izanagi's last child.

I am not retelling Izanagi's well-known story of descent into hell but, significantly, after he came back from the domain of the dead, he gave birth to his final three gods all by himself. They are born of *misogi* (ritual cleansing), which Izanagi performed after he came back from hell—he washed himself. In this ritual he gave birth to 14 gods in total, of which the last 3 were the most important and played decisive roles afterwards. When Izanagi washed his left eye, Amaterasu oo mi kami was born. From the right eye, Tsukuyomi no mikoto was born. From the nose, Takehaya susanoo no mikoto was born.

These three were particularly important because they were ordered to reign the human world as we know it. Moreover, they were endowed in many ways with human characteristics, leading to many scenes of conflict that are not quite suitable for gods' dignity.

By Izanagi's order, the oldest of the three, Amaterasu, now ruled Takamanohara (Heavenly Field), which was at the same time the daytime world. She was the sun goddess. Tsukuyomi, presumably a male and a younger brother to Amaterasu, took on the domain of the night. His name literally means "Moon Reader," and he was the god of the moon. Then the youngest male, Susanoo, became the ruler of the ocean and sea. This was the troublesome one, a trickster par excellence, who despite his physical presence and power, never grew up mentally and often threw terrible tantrums like a little child. When he cried like a baby, he made all the green trees and grasses wilted and brown, the lively seas and rivers were dried up, bad spirits went rampant, and their fly-like noises filled the air, inviting all bad things to happen at once. Then came the episodes of sibling rivalry between Amaterasu and Susanoo, both of them engaged in the god-making contest and in the ensuing disputes and fracas. Again, I am not going to tell the rather interesting episodes of Amaterasu hiding in a grotto, thus causing the total darkness of the world, etc. Eventually, their conflict was settled and eight million gods gathered and agreed to punish Susanoo for his defiant acts against the supreme goddess Amaterasu. Susanoo was ordered to offer 1,000 items to compensate for his bad deeds; then his beards and whiskers were all pulled out as well as all his nails before he was expelled from Takamanohara (Heavenly Field).

Before leaving heaven and beginning his errancy in this world on the surface of the earth, Susanoo asked for provisions for the journey. Here, our Oogetsuhime came into play. As the manager of all foods, she prepared a splendid feast of foods for Susanoo to carry with him. The hidden fact is that she obtained materials for cooking from her own nose, mouth, and anus. Curious about how she collected and prepared foods, Susanoo sneaked a look at her in the kitchen and found out that she was feeding him all those liquids and solid stuff from her own body and these, by their secretive nature, had to be unclean. Enraged, Susanoo killed her instantly.

Thus, the killing of the goddess of foods was done. And this sacrifice took fecundity to another level. The murdered body of the goddess produced the following important items that were commonly used. From her head silkworms were born. From her eyes rice grain was born. From her two ears, the millet. From her nose, red beans. From her genitals, the wheat. From her anus, soybeans. Seeing this happen, Kamimusubioya, the god of the ancestors of cereals, gathered the five items to be used later as seeds.

The case of Oogetsuhime as the foundational goddess of grain may be considered the culmination of the logic of the death–life continuum in Japanese mythology. A dead body becomes a source of life. Every body part can be regenerated, or recycled, as a god and its pertaining item. In those myths, there is no clear distinction between ideas and materials, gods and materials. All gods are natural gods, with their places and positions in the natural world and, in some cases, in some immaterial topology of things. No realistic endorsement is possible. However, the sense of life and vivacity thrives on with bits of myths, or mini-stories, serving as vehicles to convey knowledge of constructed authenticity. Of course, this sense of continuity between death and life cannot be limited to Japan. It must be widely held among land-based societies all over the planet, because when one observes the level of material circulation, the continuity is, quite simply, a fact. We will explore this continuity in the following sections.

Decomposition and life

Dead matter becomes the source of life. This fundamentally sacrificial structure may be a commonplace in world mythologies. Yet by closely looking at the process of decomposition, we are offered a renewed vision on the continuity of life at many different levels.

In the history of Japanese Buddhist painting, there is a traditional genre called *kusoozu*, or Drawings of the Nine Aspects, which depict the stages of decomposition of dead human bodies (Yamamoto and Nishiyama 2009). Such drawings exist across periods and have been attempted by many painters. Nine stages of the decomposition of the corpse are usually stated as follows: the decomposition generates gas inside and inflates the body; the decomposition progresses and the skin begins to tear and break; the corpse is very badly damaged and the fat, blood, and bodily fluids ooze out; the corpse itself loses its contour; the color of the corpse turns blue-black; maggots spring up and the dead body is eaten by birds and animals; the body parts are scattered all over the place; all flesh, blood, and fat are gone and the bones alone are left; when the bones are cremated, only ashes are left behind.

The contemplation of this process of transformation is itself a necessary training for a monk, a lesson on the vanity of this world. Our life is transient and nothing is constant—such is a traditional view commonly held among many Japanese. *Akirame* is a Japanese word that means both "giving up" and "seeing the hidden nature of things." This attitude of detachment from the current world of troubles and sadness is a constant in the history of the Japanese mentality. The famous essay *Hojoki* (1212?) and the master haiku poet Basho's travel journal *Oku no hosomichi* (1702) testify to such an attitude being deeply imbued with popularized Buddhist philosophy.

But all of these examples only look at human bodies. Possibly inspired by this tradition, the superb animal photographer Manabu Miyazaki (b. 1949) has taken a series of decomposing wild animal bodies, showing how each death feeds smaller animals and the dead body gets spread over the wide terrain of materiality. His book of photography *Shi: Death in Nature* (Miyazaki

1994) touches upon the most basic condition of our lives, "us" here being all the living. There is nothing metaphysical or imaginary in this. It is a material fact that the various components of life are constantly recycled and recombined and begin new lives under different appearances and combinations. From this point of view, we are less and less sure about the boundaries that separate not only life and death but also humans and animals, or even different animals and plants and fungi and minerals. On this planet we share the same materiality of elements. Things flow through us, connecting us beyond each individuality. Such a view of material and dynamic chains of being is at the basis of any ecological thinking. This is our main subject in this chapter, and we now begin to discuss how contemporary art can be a philosophical praxis that invites us to rethink our living conditions.

The art of Maki Ohkojima

Continuity of life and death can offer a very powerful ground for artistic creation. Examples are rife in today's ecologically conscious art. Ecology is a field of thinking in which everything is related to everything. In fact, many instances of divisions, classifications, separations are questioned and reconsidered in a broader interconnectedness of things. I will take up the case of a young, innovative, and powerfully ecocritical Japanese artist.

Maki Ohkojima (b. 1987) is one of the foremost artists in this direction. She takes the world of the living as the theme of her works, and her series of paintings, drawings, ceramics, and multimedia installations are titled *Re-forming, Mimesis, Continuous Contours*, and such. In her works, humans and animals merge—animals and plants, too. They are all in the process of constant reorganization and transformation. Anthropomorphic and zoomorphic strange shapes and colors appear everywhere, not declaring their own identity but with their transient and fluid nonidentity.

Let us take a look at one of her latest paintings, titled *Ena* (see Figures 8.1 and 8.2).

"Ena" is the Japanese word meaning placenta. But the word's connotation may be quite different from its counterpart in English. It is not a medical term but a traditionally used word that englobes all that comes out of the mother's body with the newborn—placenta, umbilical cord, and all that enveloped the fetus in its growth within. Hence the name: "cloth to cover the baby." It has long belonged to the vocabulary of midwives, and ena itself has been treated with care. Especially, the baby's umbilical cord is often dried and preserved.

Ohkojima's *Ena* is a huge painting, measuring 12 meters in width, that depicts just about everything. Seemingly, it is a forest with many different plants, animals, and fungi at many different stages of evolutionary history. As ontogeny repeats phylogeny, each individual image that appears in this painting has its own right to represent the whole history and story of its

FIGURE 8.1 Maki Ohkojima, *Ena*, 2022. Courtesy of the artist. Photograph by Ken Kato. https://ohkojima.com/#1.

FIGURE 8.2 Maki Ohkojima, *Ena* (detail), 2022. Courtesy of the artist.

species. It is a forest of many things and the matrix of life in the artist's vision. Here is the artist's statement accompanying the piece:

> Decomposition and generation.
>
> The contour of our body is forever fluctuating on the spirals of these two concurrent processes.
>
> In this work, I took my former work *The Score of Genes* (2019), to which I consigned images of a dead body in decomposition, as a seed for creation. Starting from there, this drawing began as the images of germination of many things in the process of becoming, resulting in a huge tree and the forest that envelops the tree.
>
> This museum is enveloped by the forest in a new green of spring.
>
> The process of our eating, digesting, and excreting is itself a duplication of the forest's eating, digesting, and excreting us.
>
> There is no end to decomposition and generation.
>
> All is enveloped by the forest as an immense *ena*.
>
> The forest envelops me.
>
> I envelop the forest.
>
> I eat the forest.
>
> The forest eats me.
>
> *(Ohkojima 2022b)*

This is a succinct statement about the work that looks chaotic from afar but is filled with precise images of lifeforms if seen closely. Trees, grasses, mushrooms, animals, insects, eyes, eyes, eyes. … It's as if a thousand eyes were perceiving you. And the feeling is exact, because the reciprocity of perception is at the heart of this work. *Esse est percipi* (to be is to be perceived). I look at the forest, the forest looks at me. By mutually perceiving, I and the forest let millions of beings exist.

Critic Yuki Masami uses the phrase *shin ra ban sho* with a twist. The Japanese phrase (presumably loaned from Chinese) is usually written with four Chinese characters meaning "forest," "juxtaposed," "ten thousand," and "figures." Yuki uses *shin* "the body" as the first character, replacing "forest," thus pointing to the interchangeability of forest and body. A museum talk that took place between Yuki and Ohkojima during the exhibition was aptly titled: "*Shin ra ban sho*—the Body in Connection with the Forest." Ohkojima repeatedly uses spiral images in her works and in *Ena*, too. Spirals are basic patterns in nature, from DNA to typhoons and galaxies. In Ohkojima's work there are trees tangled up in spirals, edible ferns and such, which, by virtue of their shapes, embody the logic of growth deployed along time. Growth with variations is best depicted as spirals. Writing on this work, *Ena*, Yuki stresses the ambiguity of death and life in and around the act of giving birth. She begins by explaining the meaning of this traditional term.

> *Ena* refers to the amniotic membrane that envelops and protects the foetus, the placenta that sends nourishment to the foetus, and other relevant parts. After the baby is born, *ena* is expulsed as afterbirth. In the old times in Japan, *ena* and the baby were considered as one and the child's state of health was related to *ena*'s state. So, it was preserved in a clay pot and buried in the ground until the child reached a certain stable age. On the other hand, in other locales, *ena* was considered impure and soon disposed of after birth. It was thus at the same time an object of prayer and something filthy. *Ena* didn't belong exclusively to sacredness or impurity. And seeing things from the point of view of birth leaves ambiguity ambiguous.
>
> *(Yuki 2022)*

Yuki tries here to grasp birth as a philosophical problem by quoting the important Japanese poet and author Morisaki Kazue (1927–2022). Yuki writes:

> My eyes were opened when reading writings by Morisaki Kazue. She was the person who questioned the view of life confined within one generation—I was born, I die—and kept writing about how life is transmitted across generations. She was keen on problematizing the act of birth. When you say life and death, you can't leave the dichotomy behind. But when you say life and death and birth, life and death cease to be conflictive. The resemblance in sound of womb and tomb becomes acceptable. Morisaki writes the following in her essay called "Giving Birth." "Even after one's life is over, others live. To me, the reliance on this fact warms up each death. I am infinitely relieved." Giving birth generates reliance on the continuity of life.
>
> *(Yuki 2022)*

Then following this statement on the ambiguity of life and death, Yuki pays attention to another aspect of ambiguity embedded in a forest:

> Nature in the first place is ambiguous. When you enter a forest and take a deep breath of the fresh air, you feel you are given new strength. This is an effect of phytoncide. Phytoncide means "killer plants." The pleasant scent of healing for humans is the substance emitted by the plants for their self-protection against harmful bacteria and insects. Healing and killing are inseparable in the forest and that's where the succession of life takes place.
>
> *(Yuki 2022)*

The forest is a continuum and, as such, it is the place we all belong to. It also becomes the only viable metaphor where all forms of life coexist without much mutual exclusion.

Correspondences and conspiration

What does Ohkojima have to say about her own work? The following statement is from her latest and ongoing exhibition (as of November 2022), held at the Chiba Municipal Museum of Art. The exhibition is titled *Correspondences* and is held within the framework of the museum's "Work-in-Progress Lab" as its ninth edition.

> When my skin feels a fresh, uplifting breeze, I sometimes can't help but think that this comfortable breeze might be a remnant of a sigh that someone has long ago emitted. The sigh that unintentionally departed from someone—whose name is unknowable even—becomes magnified while blowing among trees and grasses and flowers, dressed with natural aromas of the land and sea while travelling, then soared into the sky with migrating birds, finally touching my skin now, quite accidentally, far from the point of departure both in time and space. To imagine this makes me realize that this infinitesimally small here-now is really connected to all the epochs and all the scenes of the world across time and space. Of course, what visits me is not limited to sighs. There can be air emitted by a very ancient trilobite or natural gas that leaked from the fissure of the plates deep down the ocean. In this world, innumerable colors, scents, tastes, and touches are incessantly echoing each other without a beginning or an end. It's correspondences of things without limit. They relate us to ourselves as if the whole was a postal correspondence without a fixed sender and addressee. By the way, what was that person feeling when giving out a sigh that eventually reached me as the nicest little breeze? In this edition of the Work-in-Progress Lab, I'd like to take up such an exercise of imagination with you, too.
>
> *(Ohkojima 2022a, my translation)*

The word used in the title, "correspondence," of course refers us back to Baudelaire's famous sonnet. We might as well take a look at it here, so we can tune ourselves better into the mode of associative imagination.

Correspondances (Charles Baudelaire)
La nature est un temple où de vivants pilliers
Laissent parfois sortir de confuses paroles;
L'homme y passe à travers des forêts de symbols
Qui l'observent avec des regards familiers.
Comme de longs échos qui de loin se confondent

Dans une ténébreuse et profonde unité,
Vaste comme la nuit et comme la clarté,
Les parfums, les couleurs et les sons se répondent.
Il est des parfums frais comme des chairs d'enfants,
Doux comme les haubois, verts comme les prairies,
— Et d'autres, corrompus, riches et triomphants,
Ayant l'expansion des choses infinies,
Comme l'ambre, le musc, le benjoin et l'encens,
Qui chantent les transports de l'esprit et des sens.
(Baudelaire 1961, 11)

Correspondences (my translation)
Nature is a temple where living pillars
Sometimes speak confused words;
Man passes though it by way of forests of symbols
That observe him with familiar gazes.
Like long echoes from afar mixing
In a dark and tenebrous unity,
Vast as the night and as daylight,
Perfumes, colors, and sounds respond to one another.
There are perfumes as fresh as the flesh of children,
Sweet like oboes, green like grass fields,
—and others, corrupted, rich and triumphant,
Having the expansion of infinite things,
Such as amber, musk, benzoin, and incense,
That sing the soaring of the spirit and senses.

Simply put, Baudelaire's correspondence refers to the experience of synesthesia. The objects in nature per se are not thrown into chaotic fluidity. But it is at the same time a Rimbaldian systematic derangement of the senses that allows one to perceive reality in a completely different light. I am not saying there are solid realities and fluid realities. On the contrary, realities themselves are always a wide web of interpretive mesh works that are deployed across the surface of materiality. Biosemiosis works at many different levels. From molecules to organs, from individuals to species, different levels of living units constantly interpret their surroundings to optimize their survival. If we say a "derangement of the senses," one might think it is something separated from the corresponding reality, but it is not. We are talking here about a self-critical, alternative composition of reality in which many of the conventional demarcations are cancelled.

Ohkojima uses the word "correspondences" to designate not only the Baudelairean synesthetic experience but also the ecological connectivity of the material world. Not unlike Bruno Latour's actor–network theory (see Kubo 2019), each thing in the world can be an actant to perceive and act upon others so that eventually all is addressing all in one way or the other, all things "writing" to all things, making connections among themselves—ourselves—as "symbols" (remember that the word etymologically means "throwing together"). There arises another forest of symbols. The artist herself works as an individual human, but as a human individual she is never an in-dividual (undividable) unity; she, as a creative agent, responds at once at many different levels to things of this world by multiplying herself and constantly throwing together a horizon held in common. What horizon? The horizon of life as it is.

Life as com-position

What if we consider life as "being there together"? Sharing the same position will give rise to life. When two or more things overlap, the togetherness triggers a new configuration of materiality. That may be the first step to drive life into existence.

In my 2020 talk with Maki Ohkojima and the marine biologist Atsushi Tsuda, I initiated the discussion by explaining an experiment on the birth of life on Earth (Ohkojima et al. 2021). In the 1950s, American scientist Stanley Miller and his group famously made a soup-like liquid of polymer compounds and threw a series of sparks onto its surface. When it was repeatedly struck by electric charges, at one point a membrane appeared on the surface of the soup. The membrane is made of amino acids, and it was the beginning of unicellular organism—the birth of life. This experiment may be classic, but it is still very charming, because it shows that life began as a "coming together" of things in the same position. Com-position. When material conditions are ripe, a sudden strike of lightning is enough to start life.

Maki Ohkojima has been using the "soup" metaphor for some time. One of the central experiences for her was a research trip on board the schooner *Tara* that she was invited to join as a resident artist (see Wang 2019). On this trip she had daily conversations with marine biologists and experienced immense inspirations that later shaped her basic view of life on this planet and the other-than-human species. According to Ohkojima, her first revelation came when she was traveling aboard the sailing laboratory schooner and encountered the dead body of a whale in the middle of the ocean. It is extremely rare for a ship to find a dead whale, and this led her to think about the process of decomposition of life, the individuality of death, and the hidden multiplicity within. The reason why seeing a dead whale is so rare is that only when it is inflated by gas from inside does it stay on the surface. Soon it gets eaten by various animals and planktons and sinks into the abyss. Even veteran sailors rarely see one. And this was a white whale! It was a next to impossible encounter. Being there in the same spot is itself almost miraculous. Encountering a dead whale on her first navigation was a case of com-position, a sharing of the same position. Looking at the dead whale and watching marine biologists work to record the specimen, Ohkojima was led to ponder on essential aspects of life, like the Buddhist paintings of the nine phases of decomposition.

In the first place, we need to keep in mind that whales eat a lot. A whale eats millions of krill a day. When the whale dies, millions of individuals belonging to different species eat it up. Watching the process, Ohkojima envisions how much life is melted and mingled in the ocean. Hence the expression "soup." The ocean is literally a thick biological soup. So many animals feeding on the whale testify to the fact that a whale has eaten and consumed so much in its life. The sheer number of these scavenging animals, large or small, is stupendous. The number is not astronomical but biological, which may mean more than astronomical. The problem of biomass makes us think. As a species, humans are now the heaviest biomass on earth. Whales do consume a lot of krill, but we seldom realize how much life we consume every day. The amount is ridiculous. One Japanese human consumes as much energy as a tyrannosaurus rex, day in, day out. Needless to say, it is impossible to continue with this much consumption for long. The rich soup of the ocean is now threatened by humans' own activities and excessive consumption patterns.

According to the marine biologist Atsushi Tsuda, who has spent more than 1,000 days on the ocean for research and never seen a dead whale floating, whales live for 30 to 50 years but, when dead, they are afloat only for five days. If you calculate from these figures, the possibility to see a dead whale is one in 3,000 to 4,000 cases of witnessing whales. Ohkojima says the *Tara*'s captain noticed the body because there was a frenzy of gathering birds on the sea surface. It was breathtaking. One death feeds many lives. Such sacrificial instances of nature are constantly played and replayed, over and over again into eternity. But one too often forgets it when one lives in the general atmosphere of contemporary urban consumer culture. A consumer society is taken out of nature's circular temporality and thrown into the simplistic linear time structure of profits and losses. Artwork such as Ohkojima's becomes a reminder that life matters and life (the ultimate profit beyond figures) is based on death (the ultimate loss and destruction) and that there is circularity and symmetry in the exchange of life and death.

Art beyond the Anthropocene

The sense of death underlies all mythology and artistic creation, the sense of loss, tinged with an overtone of mass destruction, too. We are currently living an era of mass extinction of mega-fauna. And there is no denying that the cause of this extinction is human civilization and its accelerated modernity. Essayist Diane Ackerman at one time pointed out that around the year 1000, humans and their domesticated animals represented only 2 percent of the biomass of all mammals (Ackerman 2014). But today, the figure has reached more than 90 percent. This means that in today's world only humans and animals sanctioned by humans are allowed to survive. The land as a whole has become a garden and a hell for animals by virtue of human action. But what about the ocean? I recall a passage from Henry David Thoreau's 1865 essay *Cape Cod*:

> The ocean is a wilderness reaching round the globe, wilder than a Bengal jungle, and fuller of monsters, washing the very wharves of our cities and the gardens of our sea-side residences. Serpents, bears, hyenas, tigers, rapidly vanish as civilization advances, but the most populous and civilized city cannot scare a shark far from its wharves.
>
> *(Thoreau 1987, 219–220)*

The ocean in the 19th century was the wildest wilderness. So it is, even today, yet the sea has also gone under a tremendous change in terms of its biological potential. Again from Tsuda, the number of whales are on the rise because of the severe restriction of whaling but, as a whole, sea creatures are harshly threatened. Eighty to 90 percent of all fish species are drastically reduced in number. So are coral reefs. The southern islands of Okinawa, Japan, have the largest coral reefs in the world, but 60 percent of it is already lost due to the rise of water temperature by global warming. Ohkojima, as an artist who is fascinated by the coral reef, vividly responds to this:

> The coral reef is made by coral polyps. They are small animals but totally dependent on the plants and can only survive by symbiosis. Without algae on the surface, the polyps can't receive the energy from photosynthesis and they turn white. This is coral bleaching. The algae are like skins for polyps and when these skins are gone, they become as white as bones. I can't help associating this with the destruction of the ozone layer. The ozone layer is like skins for the earth. We need to imagine the entire earth as if it was our own body.
>
> *(Ohkojima et al. 2021, 34)*

In fact, this kind of sudden shift of scales in imagination is characteristic of Ohkojima. From micro-cosmos to macro-cosmos, from fungi to plants, from bacteria to large animals, her focus shifts incessantly. That is her free phenomenology (her perception of the phenomena). In her current exhibition, *Correspondences*, she is holding a series of very interesting dialogues (Ohkojima 2022a). She invites six guests and discusses the problems caused by humans. Her guests are Mountain, Monkey, Slime Molds, Corals, Excrement, and Body. All of them are nonhuman, the other of the human, speechless, and mostly uncommunicable. So, she asks six human agents to speak for them: an anthropologist, a monkey sociologist, an intellectual historian, a marine biologist, an outdoor pursuit specialist, and a dancer. It is a series of conceptual performances, so to speak, and each topic relies heavily on the guest's ability to present the unpresentable, or represent the unrepresentable, but this can be very interesting and eye-opening for the audience.

Among the choice of her guests, I was especially attracted by corals and slime molds. Corals offer a clear case of symbiosis. They are colonies of identical polyps connected by an exoskeleton, themselves animals and host microalgae that produce energy through photosynthesis. But when the water temperature rises, stressed, polyps expel zooxanthellae that have been in symbiosis with them and giving the colors typical of corals, leading to the bleaching and eventually to their white deaths.

One can easily imagine why corals attract Ohkojima's attention. It is because they are leading a life on the border between animals and plants and their life is hanging in the balance in this era of (most probably human-caused) global warming. Similarly, slime molds, too, give food for thought to the artist. The slime mold is rather well known in Japan because it was the main field of research for the great Japanese naturalist and encyclopedic polymath Kumagusu Minakata (1867–1941). Contrary to what people tend to believe, they are not fungi and so many different species of them do not even belong to the same biological category. They are unrelated eukaryotic organisms but share the same lifestyle of shifting between a free-living single-celled stage and the formation of spores (see Wikipedia 2023). Between these two types of life forms, each species of slime mold chooses the best for their survival according to the environment. They shift their modes and adapt to the given condition.

In a presentation given within the series of dialogues above, intellectual historian Taisuke Karasawa (b. 1978) spoke in lieu of slime molds about how they transform and move about in response to the changes around them. Brilliantly, he used many unintelligible neologisms, onomatopoeia, and even accompanying improvisation of the Theremin's strange sounds to give us an idea of what it is like to be slime molds! Kumagusu Minakata once said that by examining slime molds through a microscope one sees the macrocosm, the universe. Following the findings by Minakata, Karasawa talks about a living unit without a centre that repeats meeting and parting, responding in the optimal manner to the surroundings. In the lives of slime molds, one is many, many is one; there is in fact no unit; the border is blurred, or always negotiated. According to Karasawa, when two groups of slime molds meet, they remain motionless for hours before taking action. It is exactly as if they were thinking. Then sometimes they merge, sometimes they part ways. And merging can occur between two different species of slime molds. Interspecific merger. And during that moment of pondering upon the action to be taken, it is as if they were emitting something transparent and that something was making a judgment on the spot; in other words, "thinking" (see also Karasawa 2015).

Of course, this "thinking" is not used in the human sense of the word. There is no selfhood. They only exist as a collective. To this, Ohkojima responds by saying we should all give up the I of the human and retrieve, if it be possible at all, the I of the eukaryote, which, after all, is much, much older than any of the vertebrae and is a survivor from a much earlier stage of life on earth. For slime molds there is no life and death in the anthropomorphic sense. No human criteria between life and death work for them.

Conclusion: mythmaking today

By now I believe Ohkojima's broad vision as an artist is clear. She has been working in paintings and drawings, videos, and multimedia installations. In recent years she has been making more and more ceramic works that represent images that appear in her paintings in a more tactile and life-like way (see Figure 8.3). So, typically, her recent exhibitions show her paintings and her ceramics at the same time. Humans' psychosomatic reactions to a three-dimensional object are totally different from that to a two-dimensional image. That is a basic condition. When combined, the objects and images make a more memorable amalgam with a more evocative power. Humans and other animals merge. Animals, plants, and fungi merge. All of them follow in unison one basic working principle of the artist: death is the source of life.

FIGURE 8.3 Maki Ohkojima, *Correspondences* (detail), multimedia installation, 2022. Courtesy of the artist.

In other versions of Japanese mythology, Oogetsuhime, whom we discussed earlier, appears as a different god: Ukemochi. About Oogetsuhime, we followed the description of *Kojiki*, but *Nihon Shoki* narrates a similar story with different names and roles.

One day Amaterasu (the supreme sun goddess) sent her younger brother Tsukuyomi (the moon god) to earth to visit the god Ukemochi, who carried food in his body. This visitor made Ukemochi very happy and he abundantly vomited rice, fish, birds, animal meat, etc., and prepared a splendid feast on the table. This enraged Tsukuyomi, who shouted, "How dare you feed me with such filthy vomit?" and instantly killed Ukemochi with his sword.

Back in heaven, Tsukuyomi reported the incident to Amaterasu. She became very angry at her brother's rudeness and told him that she would never see his face again. This separated daytime and night. After this, Amaterasu sent a god named Amano kumahito (the name can be interpreted as "Bear Man of Heaven") to the surface of the earth to see what it was like. Amano kumahito found many things sprouting from the dead body of Ukemochi—five grains, silkworms, cows, and horses—and brought them back to heaven. This made Amaterasu very happy. She said, "All of these will be foods for humans" and immediately gave an order to make rice paddies in heaven and initiated agriculture.

The coexistence of these two versions—one with Oogetsuhime and the other with Ukemochi—is itself meaningful, because the beauty of myth is that no version can claim to be the final one; there can always be another retelling from a different angle and in a different setting. In other words, the plurality of versions opens the myth to the future, the time yet to come. And that is the domain of artistic creation.

These versions of myth are also strikingly similar to the myth of Hainuwele in Indonesia in which the dead and fragmented body of the strange girl Hainuwele becomes the source of different kinds of potatoes (Yoshida and Furukawa 1996). This myth of fecundity by death may have a wider resonance in the myths of the other geographical areas of the world, and it may well touch upon a basic logic of the variety of life that humans as a species have perceived. And Ohkojima's art is based instinctively on that logic. Her ideas have a lot of resonance with those of philosopher Emanuele Coccia. As I understand it, they both think that everything is in the constant process of metamorphosis. As we have seen in the case of two (and more) species of slime molds, different forms and worlds meet on the single line of—simply—life. Metamorphosis takes place between radically different species and aligns them in a shared future. In the words of Emanuele Coccia, for example:

> La métamorphose existe parce que tout vivant se retrouve à passer, dans une même ligne de vie, par les expériences et les mondes les plus divers: elle est un couloir qui permet au vivant de ne pas être obligé de vivre plusieurs vies simultanément et aux deux de cohabiter sans se fondre entièrement.
>
> *(Coccia 2020, 81)*

> Metamorphosis exists because all living things go through the most diverse experiences and worlds along the same line of life. Metamorphosis is the corridor which allows living things not to be obliged to live several lives simultaneously and it also allows two living things to live together without totally melting into one.
>
> *(My translation)*

Being one, however precarious that one may be, and that goes on living in its synchronization with the world-environment by constantly recombining one's *one*—such is the aim of life on this planet and art as a synecdoche of life. Maki Ohkojima, herself as a constant creative process in metamorphosis, incarnates and performs such art. To conclude, let us observe the characters with which her name is written. Her family name has three characters: big-little-island; her first name has two: true-tree. This semantic resonance is astounding. She is the artist who grows her own single true tree on her big, small island to bring into being an infinite forest of multiplicity.

References

Ackerman, Diane. 2014. *The Human Age: The World Shaped by Us.* New York: Norton.

Baudelaire, Charles. 1961. *Œuvres complètes.* Paris: Gallimard.

Coccia, Emanuele. 2020. *Métamorphoses.* Paris: Payot & Rivages.

Fukunaga, Takehiko. 2003. *Kojiki: A Modern Translation.* Tokyo: Kawade bunko.

Karasawa, Taisuke. 2015. *Minakata Kumagusu: Nihon jin no kanosei no kyokugen* [At the limit of the possibility of a Japanese]. Tokyo: Chuko shinsho.

Kubo, Akinori. 2019. *Bruno Latour no torisetsu* [A manual for Bruno Latour]. Tokyo: Getsuyosha.

Miyazaki, Manabu. 1994. *Shi: Death in Nature.* Tokyo: Heibonsha.

Nakamura, Hironobu, ed. 2009. *Kojiki.* Tokyo: Kadokawa sophia bunko.

Ohkojima Maki. 2022a. "Artist's Message." In the exhibition *Correspondences.* The Chiba Municipal Museum of Art.

Ohkojima, Maki. 2022b. "Watashi ga kurau, mori ga kurau" [I eat, the forest eats]. Written for the exhibition *Jitsuzuki no rinkaku* (Continuous Contours). The Saison Museum of Contemporary Art. In her home page: https://ohkojima.com/works/works-1070#3.

Ohkojima, Maki, Keijiro Suga, and Atsushi Tsuda. 2021. "Umi to tsuchi ga majiwaru basho" [The place where the sea and the land are mingled]. *Tagui* 3: 32–46.

Okuno, Katsumi. 2022. *Karamariau seimei* [Life that is interwoven]. Tokyo: Aki Shobo.

Okuno, Katsumi, Shiaki Kondo, and Natasha Fijn, eds. 2021. *More than Human: Multi-species jinruigaku to kankyo jinmongaku* [Multi-species anthropology and environmental humanities]. Tokyo: Ibunsha.

Thoreau, Henry David. 1987. *Cape Cod.* New York: Penguin Books.

Wang, Jie. 2019. "Tara's Ecological 'Deep Blue' Expedition." https://www.shine.cn/feature/art-culture/1909282877/.

Wikipedia. 2023. "Slime mold." Wikimedia Foundation. Last modified February 16, 2023, 08:14. https://en.wikipedia.org/wiki/Slime_mold.

Yamaguchi, Mikako. 2014. *Herajika no okurimono* [The gift from the moose]. Tokyo: Shunpusha.

Yamamoto, Satomi, and Mika Nishiyama. 2009. *Kuso zu shiryo shusei* [Collected documents on the illustration of the nine phases of death]. Tokyo: Iwata shoin.

Yoshida, Atsuhiko, and Noriko Furukawa. 1996. *Nihon no shinwa densetsu* [The myths and legends in Japan]. Tokyo: Seidosha.

Yuki, Masami. 2022. "Jinshinsei no ena" [Ena in the Anthropocene]. Written for Maki Ohkojima's home page. https://ohkojima.com/en/works/ena#2.

9

CODA

A radical evocation of seed

Jeanette Hart-Mann

There are moments when I sense that I have forgotten much … when distant memories grow from an amnesiac lapse, a glitch, stimulated by vegetal stories of lives that have long been systematically subordinated and instrumentalized as exploitable materials and bodies. Like a series of coalescing imaginaries crossing over perceptual thresholds, these moments give pause to what I thought I knew while creating a place for different awareness to emerge. As ecological experiences, they germinate like seeds grown plump in contact with soil and water, ready to burst forth into queer new lives of sensing, knowing, and being.

My work as an artist is rooted in these profound thresholds, though not as an abstract concept, philosophical treatise, or esthetic pursuit but rather as an embodiment of creative transdisciplinarity, labor, and love. Responding to the overwhelming plethora of anthropogenic environmental disasters driven by the tragedy of human exceptionalism, my work grows from deep histories of grassroots resistance yet strives to do more than struggle—it demands to imagine and cultivate futurities of ecologic healing and hope through seeding resilience. My art practice stems from my life as a working farmer, providing nourishment for my local community while engaging agroecology as a critical and creative eco-social art praxis (Freire 2000).

Agroecology is a scientific study, land-based practice, and social movement whose aim is to generate sustainable food systems while nurturing diversified communities of ecological life and making heterogenous togetherness possible (Gliessman 2015). I refer to my engagement as "eco-social praxis" because, for me, praxis includes both the human and "more-than-human" communities I create alongside and who have as much at stake in this as I do.[1] It is only through engaging with one another and nurturing diverse multispecies ecologies that all of our worlds might survive and thrive (see Figure 9.1).

DOI: 10.4324/9781003348535-13

FIGURE 9.1 SeedBroadcast in collaboration with Acoma Ancestral Lands Farm Corp, *Seed: Climate Change Resilience, It's Not by Chance at All*, 2019, digital photo montage, 168" × 144". Image courtesy of SeedBroadcast.

Situated in the specific bioregional contexts of the American Southwest, my practice is committed to the local and to entangled intimate relationships through daily farming, an embodied laborious process that enfolds nuances of ecologic life that may be imperceptible to other people. My practice is also durational and generative, informed by temporality as intergenerational enmeshment circulates through time and as diverse lives of biotic and abiotic expressions are revealed through seasons, years, and generations. I began this work in 2007 with the intention to continue it through my lifetime and pass it on to others and with the awareness that the local is an ethic. For me, committing to the local is about caregiving, impact, and awareness that every action has consequences. It is not about a position of good or evil; rather, at its core is the question of "who for?"[2] Who do I care for, who is impacted, and what are the consequences of my work? I want to intentionally attend to who and what, knowing that the local exponentially amplifies action on a global scale.

I grow, tend, and comingle with the complex "many lives" of soil, plants, and animals who are at home on my family farm in Anton Chico, New Mexico. For me, farming is not genteel or wielded upon the backs of others like the mastery of colonial plantations; rather, it is a part of my body's labor. Farming is hands-on, feral, collective, generative, responsive, physical, and emotional, as well as soothing, tender, and empathic; it is messy and involves the creative agency of many ecological selves. Farming is a visceral process that turns my perception to the rhythms of daily, yearly, and generational life cycles, connecting me with more-than-human worlds.

I find resonance in what Donna Haraway calls "sympoiesis," or "making with," where our many lives are entangled "through semiotic material involution" (Haraway 2016, 58). In this making with, a dialogic intersubjectivity becomes possible, stimulating reflections from deep within my ways of knowing, transforming what I thought I knew of reality, and bringing forth transfixed memories and creative new forms of knowledge. It is a momentary glimpse, thought and vision, dialogue with, and stories of the many lives who I am making with on the farm. These storytellers have become central to my work and have grounded my commitment to nurturing more-than-human stories as radical evocations to acknowledge and respond to our planetary crisis.

Unprecedented numbers of flora and fauna have perished over the last 100 years due to anthropogenic changes in the earth's biosphere. Based on a recent Intergovernmental Science-Policy Platform on Biodiversity and Ecosystem Services Report and what scientists are now calling the sixth extinction, "Nature is declining globally at rates unprecedented in human history. [...] Around one million species are threatened with extinction, many within decades [...]" (IPBES 2019). Yet plants and seeds continue to teach us humans how to respond ecologically; they show us how to learn and live in the world based on eco-social relationships. Seeds grow in their own unique ways, and each seed has a story to tell; they are more-than-human sensing bodies, transformative and evocative, who carry deep memories of how to regenerate life in an uncertain future. Seed stories have become some of the most important aspects of my work—listening, making with, and giving them a place to thrive and grow.

Stories are radical like roots, undermining what we think we know and wielding us topsy-turvy within the earth beneath our feet. Stories encourage us to connect with and spread ourselves in strange new ways; they give us moments to pause, wonder, and remember while also testing our beliefs. Stories are radical as discursive ecologies, spreading far and wide as generative, diverse expressions. Alive and vital to all communities as they root and grow, stories can be grafted together across lives, environments, and histories, creating chimeras of polyvocal perception that allow us glimpses beyond the solitary realities that we think we know.

The word "radical" is doubly rooted, for its Latin origin is *radix* or "root." Radix is also the origin story of a seed. As the embryonic root that first emerges, the radicle is not merely a mechanism by which a seed becomes a plant but is rather a bodily organ, a physical expression of complex agency and desire prying through the seed coat, reaching forth to nourishment, and initiating a living relationship with other ecological selves. For more than 100 million years, seed-bearing plants have undertaken this simple act with profound consequences (Friis et al. 2011). As seeds root to become plants and plants flower to become seeds, they accumulate experiential knowledge of this ecological context and likewise put this to use in their vegetal anarchitecture (Marder 2016).

Because of eco-social relationships, plants embody forms of remembering that their seeds carry with them over generations. The embedded information stimulates plants to make with their environmental context in order to survive, enabling them to flourish seasonally, yearly, and generationally, but it can also be a reminder of what they do not know when confronted with new environmental pressures and challenges. I cannot help but feel kinship with plants and seeds and to anthropomorphize this moment; it is like an uncanny coincidence to wonder, dream, question, and fear for the unknown and then feel inspired to do something about it. What do plants do with this information? They change. They adapt, mutate, diversify, and find ways to live in the trouble (Haraway 2016), or they die and disappear forever.

Radical is seed as resistance. Like other more-than-human storytellers, seeds are revolutionaries. The stories seeds tell are humbling and direct, refusing the false hope of business as usual and the Anthropocene's imperium. Their stories reveal the ecocidal consequences of entrenched hubris on the planet and yet express ecological resilience and ability to enact monstrous hope, giving us direction and encouraging our generational participation with plants, food, and, again, seeds, year after year. As radicals, seeds demand that we listen to their stories and tend to our responsibilities. They show us ways to create change by planting one seed at a time. They connect us to embodied ecology through cycles of sustenance while teaching us to grow our perception of animacy beyond the human. A radical evocation of seeds is a call to action—to listen deeply, bear witness, and make meaning together to respond to our collective ecological crisis.

Listening requires being with and sensing in different ways and making space for intentional presence and acute awareness; it also means prioritizing nonjudgment while making the effort to honor stories, each in their own way.

Ecological listening means shape-shifting the way we situate human perception and language to realize other forms of meaning making. I feel privileged to have a daily practice that requires me to be in proximity with this expanded field of sensual knowledge, but it is never without effort. My work requires releasing all tensions of imprinted knowing while being affected by these stories.

Being an artist is akin to being a farmer; for both, critical and creative inquiry are woven into embodied interdisciplinary labor with the intention to cultivate a kind of nourishment in the world. Art has always been a pedagogical and ecological inquiry for me, one in which making is infinitely filled with opportunities to cultivate relationships among everyday experiences, ideas, poetic processes, and making with others. For me, making art also requires a deep commitment to generative collaboration with others. Be it with land, plant, animal, or people, collaboration is one of the most challenging and transformative processes in my work. I must bring my full creative energy to making with others while being willing to have my desire, beliefs, and creative skills stretched, broken, and supported.

Positioned at the intersection of art, agriculture, activism, and environmental and social justice, my work is transdisciplinary and ecological. I employ art forms such as sculpture, installation, video, sound, and photography, as well as performative processes grown out of farming, social engagement, collective action, organizing, and writing. Each of these forms is a process of inquiry and possible material for conveying agroecological stories and augmenting my perceptual awareness. These processes have been especially useful for me in bridging the distance between the intensive goal-driven labor of growing plants to successfully produce food and my desire to let intuition and curiosity guide my approach to creative research on the farm. Over the last 15 years, I have employed photography, video, and acoustic explorations to seek out the voices of ecological life. As documentary processes, they are starting points for sensing more acutely and discovering gaps in my own perception. They are also opportunities to wonder what has been lost in translation across sensorial modes, which inevitably opens up queer revelations where I can encounter the more-than-human on their own terms.

Listening to the corn grow

I grew up in Ohio, in a region of the Corn Belt, on a small, multigenerational family farm where we, too, planted corn. The Corn Belt was (and still is) a terraformed industrial landscape in the Midwestern United States spanning over ten states where, over the course of 150 years, millions of acres of original presettlement prairie, wetlands, and cultivated Indigenous polyculture landscapes were plowed under to make way for the industrial cultivation of monolithic corn fields (Hudson 1994). In 2019, over 90 million acres of corn were cultivated with a vast majority planted in the Corn Belt (Capehart and Proper 2019).

In the 1970s, my family and neighbors talked about how they could hear the corn grow; the sound would tell them how well the crop was doing that year. In the middle of the summer, when the weather was perfect, the corn would grow by inches, day after day, standing as giant, green tropical grasses swaying and growing in the breeze. During this time of year, as the corn was thriving and settling into the flowering phase, their plant fibers would crackle. People stopped to listen as if suddenly remembering that there was something important that needed to be heard.

The sound the farmers heard was one of potential, an expression of vigorous growth and substantial grain yields, a story of bountiful harvests. As a dialogue between human and plant, I remember this exchange was matter of fact, although no one I knew would have admitted that they listened to what plants have to say. But meaning was exchanged, and the "ecology of selves" in this semiotic exchange pointed to something we humans do not typically consider (Kohn 2013, 78). Corn plants make meaning through their bodily utterances and through their situated relationship making with others. What they say, when they say it, and who listens are important.

I tried to listen to the corn on my farm in 2010 during one of the hottest years on record. I sat in the middle of the field for hours with my ear pressed against cornstalks and even went so far as to listen with an amplified audio recorder. I did not hear anything except munching grasshoppers and harmonizing cicadas in the dry summer heat. I was disappointed not to hear the corn growing and crackling, but during the time I spent in the field, the corn taught me that silence is also a sound. Silence stimulates and speaks in different ways about other kinds of futures, ones that are terrifying to live in and think about. Responding to the drought, extraordinary heat wave, and ongoing climate crisis, the corn's silence portends an ominous story. Due to another summer with higher temperatures and less water, the corn was stressed and growing very slowly, silent while regulating the life force for the little seed being produced. The corn was not crackling but dying.

Sitting with the corn, I began to wonder what I would see if I could witness this silence and "listen" by watching the corn over the remainder of the season, hour after hour and day after day. As my first crude experiment with time-lapse photography, I found a cheap solar-powered wildlife camera that I set up on the cornfield's edge and programmed to take photos sequentially from dawn to dusk. Collecting thousands of images between August and October, I waited anxiously to see what would happen. Once the corn was harvested, I sat down in front of my computer to digitally piece together the individual photographs into one moving image.

A lapse in vegetal time

Time-lapse photography captures one image after another in time intervals determined by the photographer, who typically combines the singular images into a linear frequency akin to film. Most film is shot and shown at 24 frames per second (the way humans see motion in the world), which makes the spaces between images indiscernible. With time-lapse photography, however, these gaps are opportunities to experience the lives of vibrant matter (Bennett 2010). Time is stretched in multiple directions, altering the space of the subject's movement while compressing the temporal context of our experience.

The early amateur naturalist Frank Percy Smith first developed botanical time-lapse techniques to explore flowers that bloom too slowly to see in real time. In his 35mm silent black-and-white film *The Birth of a Flower* (1910), Smith demonstrates how different flowers bloom by recording each as a short vignette. The petals unfold, grow, move, and wave across the screen in slow, steady motion. These moving images last for several minutes, showing us the birth of a flower as an animated gesture of the vegetal, queering what was idle into volatile life. When this film was first shown in the early 20th century, it captivated audiences and challenged commonly held distinctions between plants and animals, that one was active and alive while the other was passive and inert. Smith's time-lapse opened people's eyes to the life of plants and a radical new story of vegetal vitality.

Watching plants through this photographic process unveils a paradox in our human relations with the more-than-human world. Typically, we sense plants at different stages of their lives such as when their flowers are in bloom. They are the perfect subject for a singular-framed photograph because we experience them in our lives as still creatures, inanimate and unmoving. The photograph is in many ways an equal simulation to how we see plants in our lives. But time-lapse photography shows us what we miss and the fallacy of how we perceive the world. There is a gap between what we experience and what happens. We do not witness a plant's life as it is unfolding, and in this we are not aware of plants as animate. But, like us, they are living and in motion, as seed, body, root, and flowers that become seed once more.

I did not know of Smith's work when I began my own time-lapse project with corn in 2010. I cannot say why Smith did what he did, but I do know why I did. Corn is my relative, and relations are complicated. My connection with corn is filled with reciprocity and love, yet it is also coupled with past trauma. Tied to a long history of settler colonial exploitation, genocide, ecocide, and now climate crisis, the sixth extinction, and food injustice, corn has been at the center of many worlds, peoples, and events. On one hand, corn has been a mythological mother for thousands of years, making with people to ensure that generations are nourished; on the other, it has become one of industrial agriculture's most important engineered products, rationalizing logics and mythologizing legitimized master narratives and histories. These distinctions across myth and story are doppelgangers to be sure. But when interrogated through forms of embodied eco-social storytelling, my work as an artist and farmer attempts to stimulate a counterpoint where embodied story-making is an ongoing form of environmentalism and evocation of futurities.

Listening to the Corn Grow, my first attempt at sensing the life of plants using time-lapse and moving images, is a 3-minute color digital video combining sequential images of a corn field from sunrise to sunset (see Figure 9.2). It is an exploration of listening to the silence of the corn struggling through another day and another year of the engulfing climate crisis. As a story and witness to performative ecology, we watch as the corn negotiates survival by transforming back into seed once again.

As my first crude experiment in time-lapse video, the visual elements in *Listening to the Corn Grow* became cues for future artwork such as *Would We Have Seed* (2018–2019), a season-long time-lapse of my farm growing arid land–adapted crops from seeds gifted by bioregional farmers who were working to cultivate seed and food sovereignty in their communities (see Figure 9.3). Returning to the gap and the space between each time-lapse image, my interest is in what this in-between space holds: time, ecology, and sympoiesis, like the contradiction in the sound of silence filled with the ecology of selves. The gap is a part of the story that is not a representation of what is; rather, it is the actual process we are making with where we can wonder, remember, and listen more intently to the stories of more-than-human worlds.

FIGURE 9.2 Jeanette Hart-Mann, *Listening to the Corn Grow*, 2010, video still. Image courtesy of the artist.

FIGURE 9.3 SeedBroadcast, *Would We Have Seed*, 2018–2019, video still. Image courtesy of SeedBroadcast.

During the summer when I recorded the images for *Would We Have Seed*, my family and I watched as the farm suffered from drought, high temperatures, and major flooding—another year of global warming and extremes. The leaves droop by inches throughout the day and rise once again at dusk. Then, one evening at sunset in early August, the downpour begins. Within a half hour, a foot of water sheets across the fields, burying low-lying crops. All of this happens right as the camera goes dark due to the storm and, eventually, nightfall. The scene portends an ominous view, but only for a second. At sunrise, the distant view looks slightly different, a muddy blurred green but not distinct enough to recall the events of the previous night. In highlighting this part of the artwork, I am interested in how we perceive and respond to our limitations of knowing. If there is only a visual glimpse of destruction and its representation is a blur of what we think we know, will it ever trigger a response in our need to deal with climate chaos and its many rippling consequences?

Would We Have Seed was presented in 2019 at the Albuquerque Museum, in New Mexico, as a video installation in the socially engaged art project called *Seed: Climate Change Resilience*, organized by Seed-Broadcast, a collective art and "agri-Culture" project I co-founded.[3] The exhibition explored seeds, resiliency, and climate change in collaboration with land-based communities in New Mexico through visualizing their arid land agri-Culture as a series of artworks.[4] The life-sized video brought vegetal life into proximity with the viewer, deploying the phenomenon of time-lapsing to generate hyperawareness of plants growing together as a diverse mosaic of living bodies, patterns, and expressions grappling with climate change.

Would We Have Seed is both a question for and an evocation of the ecological community; it asks us as viewers to consider where all the seeds have gone while also prompting us to remember that seeds are vital collaborators as we all respond to climate change. The work draws us into the polyvocal story of seeds as they grow into living bodies attuned to the rhythm of multispecies ecologies and yet whose lives are threatened by our neglect.

As humans, how we experience and perceive the world in our daily lives and what we remember are critical in a time of crisis. I have found the process of using time-lapse as an exploratory tool insightful, albeit disconcerting. Viewing plants through this process has allowed me to wonder and delight in their animacy and taught me to see and think in radical new ways; it has deepened my understanding of story and expanded story forms into performative ecologies. The process has also given me a moment to pause and consider critical issues of time, memory, trauma, and madness as they affect our collective reality with the climate crisis. I refer to this pause as a sort of glitch, a rewiring of intersubjective perception, knowing, and making to acknowledge what we cannot see and do something about it. I see the glitch as related to the lapse or gap in time-lapse photography, where perception and memory assemble experiences of sympoietic processes that are neither here nor there but instead in the making with. These in-between spaces generate a place of experiential awareness and thought where story lives and grows between past and future.

Peter H. Kahn, Jr., professor of psychology at the University of Washington, coined the term "environmental generational amnesia" to describe how environmental degradation is made invisible through a sort of experiential time-based stupor. In this, an environment's degradation occurs piecemeal generation after generation; children are born into these contexts and experience them as normal. But as each generation grows up in ever-worsening conditions, the environment grows increasingly more degraded and yet awareness of it remains normalized. I would argue that along with this we seem to be catapulting into exponential intergenerational oblivion. It is as if our memories cannot hold on to the nearest marks of environmental disasters that we experience. As these disasters pile up and the environment continues to degrade, we do not see the crises accumulate or the rippling aftereffects; we only remember the random events that took place and see them as inconsistent echoes of some distant occurrence. But even these memories fade rapidly, as if none of it ever happened. Cognitive dissonance is also mirrored in our habitual orientation to believe the climate emergency is yet a future event at a distant location, waiting, as we wait to do anything about it.

Environmental generational amnesia and cognitive dissonance go hand in hand as we destroy ourselves along with the earth. Perhaps this is similar to what geographers call "environmental imaginaries," where the environment shapes who we are as we shape the environment. This binding fills us with ecocidal mania as we live in and make with climate chaos. How can we mend or puncture this vicious cycle? Is it a function of ripping ourselves apart to separate us from the madness and perpetual trauma? Or can we attend to this through stepping more fully into our relationship with Earth's many ecological selves to begin healing together? Is there an ecological imaginary who can cultivate and guide us on this path?

Through my work as a farmer and artist, I have come to see multispecies storytelling, making, and listening as thresholds into this entanglement so I can grapple with cycles of environmental trauma. As Amitav Ghosh suggests in his book *The Nutmeg's Curse: Parables for a Planet in Crisis*, "The planet will never come alive for you unless your songs and stories give life to all the beings, seen and unseen, that inhabit the living Earth" and, likewise, "[…] a necessary first step toward finding solutions is to find a common idiom and a shared story—a narrative of humility in which humans acknowledge their mutual dependence not just on each other but on 'all our relation'" (Ghosh 2021, 84). Like the gap or glitch, stories are living spaces of relational exchange where ecological imaginaries, like the stories of seeds, nourish our capacity for intersubjective awareness and healing. This alone will not solve all the problems we face, but it is a first step in altering blindness, amnesia, and the normalization of anthropogenic environmental violence into something we can deal with.

Seeds carry each generation's sympoietic memories, not as anachronistic artifacts but as ways of being in the present and as accumulations of living generations for the future. As kernels of potential, these memories give the seeds and the plants they become resiliency, adaptability, and creative mutability as they make with other ecological selves and confront climate chaos. Their memories are enacted through their biological life and the relationships they have with others to become stories that weave many worlds together. These stories are expressed differently from the spoken word. Yet, their language is as communicative, semiotic, and important as our own. Seed stories prompt us to expand our notion of story, the forms these take, and the lives that they bring into being. Robin Wall Kimmerer calls this "the grammar of animacy" (Kimmerer 2015, 48) If we can acknowledge the animacy of other species, then we might also value what they have to say and how they say it. As Kimmerer writes, "We don't have to figure out everything by ourselves, there are intelligences other than our own, teachers all around us" (58).

What stories will your seeds share?

Farming in the high desert of the American Southwest is a testimony to the power of seed memories, the stories they tell, and the futures they make possible. When I began farming in New Mexico in 2007 and wanted to grow food for people, I did what most farmers in the United States do—I bought seeds. Those seeds came from farmers who do not grow food but instead grow seed. Growing seed to grow seed might seem like a strange paradox—and it is—but this finely tuned industrial activity supports modern agriculture and our global food system.

Over the last 150 years, industrial agriculture has honed farming into a technocratic process of mechanization, uniformity, predictability, and massification that controls all aspects of growing food today. One way it has done this is through developing plant varieties tuned to this logic while coupling our industrial food system to match, resulting in the cultivation of minimal select food crops that make this possible and the extinction of those that do not. As a result, it is estimated that 75 percent of the world's crop diversity has been lost in the last century (UN News 2010). To get a better handle on what this means is to understand that this represents the extinction of 75 percent of plant–human relations cultivated since the beginning of agriculture, or 10,000 years of life lost in 100 years. The seeds I bought and planted struggled to live, let alone make food on my farm. But it was not just the seeds who were out of context, it was the story they told.

In 2008, my neighbors Magdalena and Cecelia started to drop off seeds for me to grow. These elderly sisters—who had beautiful and bountiful gardens situated within the arid vitality of Anton Chico, an old agricultural village—gave me seeds from flowers, medicine plants, and blue corn, old plant varieties not for sale in any seed catalog. Instead, these seeds had been saved for years by the sisters and passed from hands over generations. As I placed the seeds into the ground, I realized that I was being threaded into their stories as they were into mine. Seeds, plants, and food are gift givers of life, transforming sun, soil, and water into nourishment and familial multispecies ecologies. Seeds are situational, having learned to adapt with a genius of place that has enabled them to thrive in diverse and extreme environments through their relationships with humans, living in tandem, in situ through thousands of generations.

As the summer progressed and these locally adapted plants thrived and became seed once more, I found myself captivated by their stories. I decided to reimagine farming through a co-commitment to growing plants from seeds saved year to year. As ecological imaginaries, seeds and their gifts realize new economies of biological interactions as they exchange nutrients while making with others. Through the seedbearing process, plants feed soil, insects, other plants, animals, and people while producing large quantities of seed to grow more plants. Creating space on the farm to intentionally grow seed and food transformed my life and the life on the farm, causing major shifts in my perception, thinking, art making, and agroecological practice and teaching me that to feed people, we must also work hard to feed and care for all lives that in turn nourish these ecological cycles. Today, 15 years later and only a mere nanosecond in the storied life of each of these plants, the farm has rooted itself into a radical evocation of seed.

The gift of seed is a community responsibility that connects the more-than-human world with the everyday lives of gardeners, farmers, and everyone who eats. When Magdalena and Cecelia gave me seeds to plant, they also gave me a responsibility to continue this sharing cycle, which in turn inspired *Seed-Broadcast* and connected me with seed saving communities in New Mexico, across the continent, and globally. The seeds that now grow the farm and the food my family and I cultivate are gifts from the relationships and lives of arid land farm communities and seed stewards who saved these environmentally and culturally important plant varieties from going extinct. These plants thrive on the farm because they are at home in the high desert and they respond to the embodied caregiving I provide. As teachers, seeds share their stories and nourishment while asking little in return except that we continue the caring and gifting cycle for future generations.

Through this work, I learned about Indigenous histories and cultures that have been (and still are) displaced, violently colonized, and in many cases commodified and appropriated through genocidal policies of colonization and capitalism in the United States. Over the years, many of the seeds I was gifted were from seed savers who had no connection to Indigenous communities, yet they freely shared these seeds and their histories without reconciling this trauma. For many native communities, seeds and plants are relatives, and they are woven into sacred cosmologies. They are gifts but, more important, they are responsibilities. Seeds are family to be cared for, honored, and renewed through ceremony—to treat them in any other way is to perpetuate the legacies of settler colonialism, exploitation, and violence.

For many seed savers and growers, there is an ignorant bliss and historic romance that comes with working with seeds. Seeds emit strange magic and can carry you away with dreams of abundance. So, how do we respect and care for seeds, and how do we share their stories, especially when they are fraught with such tragedy? How do we mend the chaos catapulting us into ecological disaster when every attempt uncovers another layer of the problem? Such questions raise more questions about the embedded mythos of capitalism and colonialism in our everyday and how these infiltrate even the most benign relationships between people and the natural world.

Owning life is slavery, erasing animacy, agency, and familial connections to eco-social relations. Such ownership also threatens our ability to work across communities to ensure intentions that prioritize equity, mutual respect, and caregiving. Over the last 100 years, the privatization and commercial patenting of seeds has led to a new colonial exploit: the legal ownership of life and domination of it. Genetic engineering, or genetically modified organisms (GMOs), is the wave of the future, or so says every investor who will double their profit margins from it. But for many concerned citizens, farmers, gardeners, and cultural and environmental stewards, GMO crops, proprietary and terminator seeds, and all the chemical cocktails needed to keep them alive are a serious threat to the earth, our food systems, and the basic human right to save seeds and grow food. The hostilities feel particularly heated when it comes to corn. GMO corn is the most widely grown global crop and accounts for up to 90 percent of all corn grown in the United States (Center for Food Safety, n.d.), used to feed livestock, for ethanol, and as an artificial sweetener. Over the last decade, thousands of pounds of GMO corn seed and corn fields have been burned in protest and 32 countries worldwide have banned or severely limited GMO corn production. Many people have come to hate corn.

For years I wondered how seed stories, corn, and my life were interconnected. I wondered what a seed story could be. It seemed like a horrible contradiction to love a seed, a plant, and a food that was so important to the Native Americans my family forcibly removed through their settlement of the Ohio Territories in the early 1800s and then to follow this settler colonial story through its exploit as a catalyst for further land grabs, propriety, terraforming, commodification, hybridization, and finally genetic modification. Where was the love in that? Yet, growing up with corn everywhere in my life, playing in corn fields and giant wagons filled with seed or shucking sweet corn on a hot summer day with my grandparents, I have always felt smitten, as if corn were family.

Indigenous activist, mother, poet, seed keeper, and doula Beata Tsosie-Peña put this contradiction in perspective for me while discussing seed sovereignty, food justice, and community health. Talking about genetic engineering, industrial agriculture's destructive force, and many people's hatred of GMO corn, she pointed out that it is not the corn's fault; instead, it is the natural consequence of colonization and capitalism. If force and hate perpetuate slavery and violence, then the mission of decoloniality must be to emancipate and heal. But how?

Letter from a SeedBroadcaster (see Figure 9.4) is my attempt to mobilize my work as a farmer and artist while beginning to decolonize my own seed story, exploring the wider concept and what such activities might mean to other people. This creative work was prompted through acknowledgment that many of the bioregionally adapted seeds I was growing on my farm were not my seed legacies or origin stories to share.

My attempt began as a writing experiment and quickly took on the form of a letter. There is something deeply personal and reflective yet conversational about writing a letter by hand. The process created a humble and authentic space where I could ask questions, respond, and release memories long buried. Writing is a laborious process for me; writing by hand is this and

FIGURE 9.4 Jeanette Hart-Mann, *Letter from a SeedBroadcaster*, 2012, video still. Image courtesy of the artist.

more, as it stimulates another form of embodiment and inquiry. Without realizing, I was generating a space of creative and critical alterity, much like the gaps or glitches in my time-lapse experiments.

At the crux of the letter were my questions of lost seeds, familial connections, and what it means to cultivate generational stories tied to the traumatic past and a transformational future. Once this letter was more or less complete, I felt it was important for me to respond to it visually through exploring the farm with video and photographic camera in hand, to see whether what I was writing about was reflected in the many ecological selves on the farm. The juxtaposition of the letter, past memories, videos, and photographs from the farm grew together and became an assemblage and a call to action. What follows is an excerpt from *Letter from a SeedBroadcaster*:

> What are seed stories?
>
> Seed stories are difficult to locate as specific subjects. Like seeds, they are entangled in a meshwork of cycles and interconnections across history, pollinating the thoughts and memories of our everyday fare. They are connected to the people who enact them, yet they reach out and touch others in the most unexpected ways. Like seeds that are shared at libraries and swaps, seed stories are not necessarily stories from the past. They are stories right now. They are each of us, as we practice what it means to be twenty-first-century cultivators of vitality, through the seeds we save, share, and plant. These stories are present in the lived labor of possibility, every time we sow a seed, harvest its yield, allow space for the growth of more, and pass these seeds on to others, continuing the

movement and putting into practice an intention to make a truly viable food web in the hands of people.

As a spectrum of expressions—from the pragmatic how-to's of saving seeds and community organizing of seed libraries, to the deeply rooted, awe-inspiring, magical tale of a sprout—each and every seed story is important. These stories bring together a familial perception of the way seeds and people create communities of interchange, feeding people around the world, while creating agricultural environments to make more seed and food. Seed stories are the roots of agricultural inclinations that every human alive and throughout history has participated in. Even though we might not all save seeds or grow food, everyone eats. And, if we think about it, almost all of our foods—grains, vegetables, fruits, and nuts—are either derived from seeds or are literally a production component of the botanical seed bearing process. We all carry seed stories and we are all writing these stories right now. But, where have all the seeds gone?

I have a distinct memory of playing in giant grain wagons filled with corn and soybeans as a child.

Growing up on a small midwestern farm in the 1970s and '80s, I remember this vividly, but only now. This memory is ironic. For now, instead of frolicking around in 300 bushels of #2 yellow dent corn, bound for the feed grinder or the local elevator, I find myself holding a profound stash of corn seeds the color of earthly gems. These kernels carry with them a deep lifeline of people working together committed to feeding communities, year after year, while planting seeds of hope for an abundant and diverse harvest. Brought forward through history, as a need and desire for life, these seeds have been passed among thousands of dedicated hands, to the soil, and back again.

A half-pound of seed produces all of the corny meals for my family of four, plus enough to save and share for the next year. Cornmeal, tortillas, posole, tamales, and of course, fresh roasted corn. It also produces fodder for the goats, a trellis for beans, shade for greens, and a jungle-like habitat for a plethora of creatures.

Corn has taught me to listen very carefully to the sound of growing plants and the conversations at play in this microecology. Translating these myriad of interactions, one finds other ways to see and explore cultural logics of diversity, which counter the "agricultural sublime" of weedless, colossal, homogenized figures of green, modified over the last hundred years as the necessary companions of machines and globalization (Pollan 2001, 200).

This motley corn, a radical among cultivars, desires to be hand planted in a rowdy patch of melons and squash, in triplets per basin. Corn's seasonal life is never the same from year to year. Some years the plant is short, sometimes tall, with kernels in rows and/or squeezed into hexagonal patterns, colored blue, red, white, pink, sunburst, speckled, translucent, and opaque. Corn is never the same, corn is always changing—responding to the present conditions of soil, water, and sun, the cycles of pests and pollinators, and even the moon. Corn's genius enfolds adaptability, variability, and collectivity. While never submitting to a precise measure of character, corn is never inhibited by the standardized protocol of monomimicry and industrialization but is necessarily social and always at work within a community of others. A process striving to become seed once again and be planted the next year. This corn is an expert of probability, survival, and inquiring how-to. What else will these seeds share, if I listen to their stories?

It is through this process that I come to see myself in this narrative, for my own agency and subjectivity is implicit in how these garden dialogues evolve. These conversations affect me, as I reflect and act with the intention to grow food and also grow my thinking to engage in critical agri-Cultural practices. To plant and cultivate and spend time listening. To grow and learn. To lose control at times and allow the resulting chaos to be my creative mentor. Through accumulation, story moves me to believe that every seed and every seed saver, and every grower and lover of food, has crucial stories to share. This wealth of local, practical, and creative know-how is a site of unconditional knowledge building and the frontier of truly innovative people based, food practice.

In solidarity,

a SeedBroadcaster

P.S. What stories will your seeds share?[5]

Writing *Letter from a SeedBroadcaster* and using my camera to explore seed stories and the multispecies ecology on the farm, I naturally found myself enfolded in the story of corn. Corn is not my ancestral origin story, but I was born into and grew alongside it for the first 18 years of my life. Like me, corn has been an unwilling peon in a long and traumatic settler-colonial legacy that we were born into and carry with us, even if not by choice. However, by choice is my love of this brilliant plant—the seed, food, and figurative body and the way we seem to grow each other from year to year.

Hopeful monsters

The word "corn" has its origins derived from Proto-Indo-European roots meaning grain and, etymologically, a "worn-down" particle. As a generic descriptor of the original plant that was known as *mahiz* to the Taino people (and no doubt has many more names to millions of Indigenous peoples of the Americas), colonists new to this continent called it, along with all cereal grains, corn, and specifically Indian corn. As it spread globally, its names grew as did its story. In the English colonies, as European settlers continued to violently displace and destroy Indigenous communities, steal land, and build their own farms and villages, they continued to cultivate fields of Indian

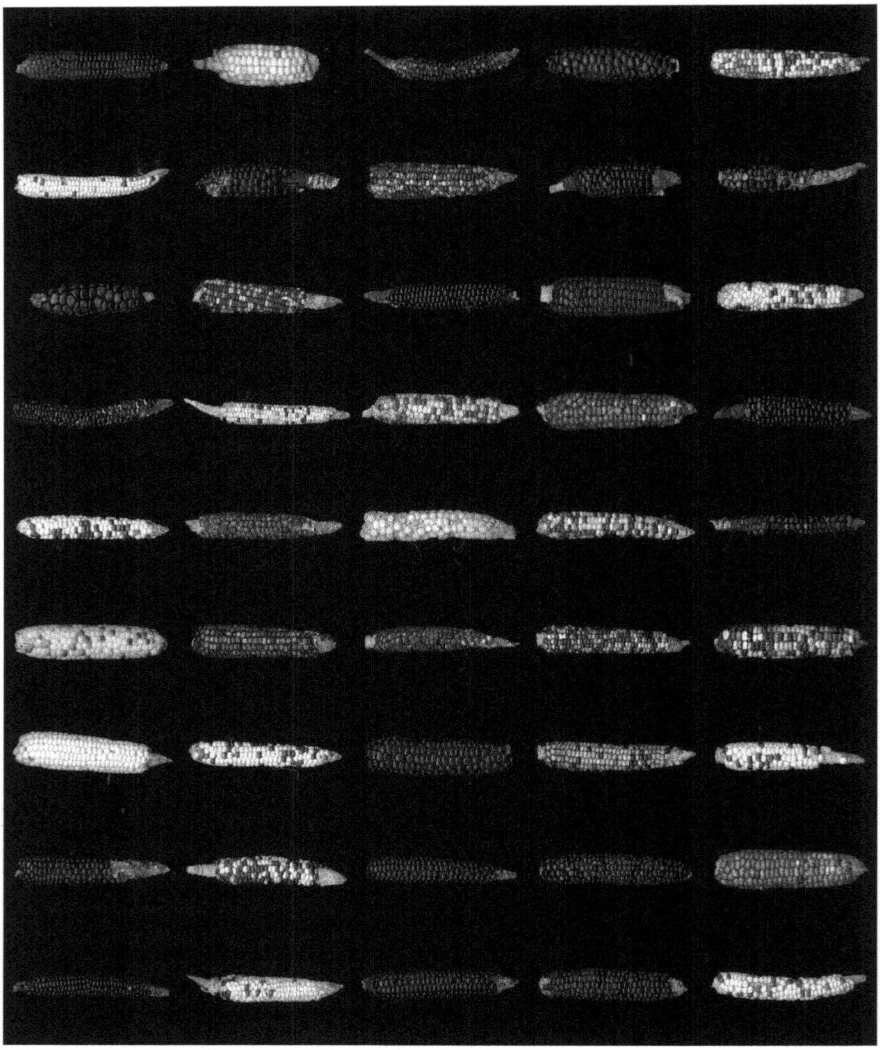

FIGURE 9.5 Jeanette Hart-Mann, *Corn Morphology*, 2010–ongoing, select digital images of archive. Image courtesy of the artist.

corn. But corn, like all colonial-era cereal grains, was classified by nutritional and culinary value in accordance with class status; at the top of this was wheat and at the bottom was Indian corn. Yet, corn's ability to thrive in almost any environment and grow faster and yield higher energy conversions from seed to plant to food and feed was harnessed as a catalyst for the expansion of livestock production, commodity markets, and the rise of the Corn Belt (Fussell 2004).

Perhaps the Corn Belt was the first large-scale system of factory farming and concentrated animal feeding operations (CAFOs) in North America where the exponentially scaled and systematized cultivation of corn supported unprecedented growth of livestock production and vice versa. Along with this, the speculative interest in commodity investment, and the centralized market distribution hub based out of Chicago, put corn squarely in the center of industrial agriculture and the rise of capitalism in the United States. As farmers continued to grow corn, they began selecting and breeding it for more production and profit. Because of its importance in feeding livestock, desirable traits included increased starches, higher yields, and uniform physiological characteristics adapted to mechanization. The development of Corn Belt dents was perfect for this, grown from the hybridization of southern dents and northern flints.

Breeding corn and creating new improved varieties is relatively easy because corn is wind pollinated, and planting different varieties in proximity ensures cross-pollination. Corn Belt dents, even though developed through a process of hybridization (crossing two different corn varieties together), are not what we would call hybrids today. Corn Belt dents were stabilized crosses and considered open-pollinated varieties with repeatable traits that could be grown and saved from year to year by farmers. This was an era when seeds were grown by farmers, not only to grow more seed but to grow crops that could be used for food, feed, and seed. The popularity of these corn varieties was glamorized through fairs and corn shows where the best-looking corn was thought to be the best to grow. As we know now, looks can be deceiving.

By the 1920s, efforts were underway between the US government, agronomists, farmers, and business interests to develop corn varieties with increased yield and vigor. At this time it was well known that hybrids tended toward these traits; this, in turn, led to what many refer to as "the miracle age of corn," which involved increased industrialization, abandonment of public plant breeding programs, farmers buying seed instead of growing it themselves, and the creation of national seed companies. Throughout the 20th century, Corn Belt dent lineages would become the hybridized relatives for the most widely produced corn in history, #2 yellow dent corn, which is in almost everything we consume, including packaged foods, meat, and even the fuel in our cars. This legacy also paved the way for later genetic engineering, patenting, and the privatization of life itself.

In 2016, as uncanny seeds go, I randomly met an agronomist who attended The Ohio State University (OSU), a preeminent public land grant university central to the commercial improvement of Corn Belt dents.[6] While he was conducting research for his Ph.D., there was a terrific tornado that took out the power to the OSU Agricultural Research Station and also the deep-freeze seed storage that held original varieties of early 20th-century Ohio Corn Belt dents. Many of these now-extinct varieties are the parents of #2 yellow dent corn, which grew today's global corn empire in league with multinational seed and chemical companies like Pioneer Hi-Bred International Seed Company (now owned by DowDuPont). These original Ohio Corn Belt dents are deeply entangled in not only the proclaimed miracle of global agriculture today but also the ecocidal mania that attends to it.

In an attempt to clean up the mess and save the seeds, the anonymous agronomist decided that liberation was in order. The fact that these seeds were very old and probably not even viable lent some credence to freedom over trash can. Having saved the seeds, he gave them to several friends and grew a plot of this motley corn himself. Growing all 17 original varieties of Ohio Corn Belt dents together, his friends and he hedged bets on which seeds would still germinate, ensuring pollination success with whatever did grow. Through these efforts, he discovered that some of the seeds were dead but others were still viable. The seeds that did grow restored now-extinct and threatened varieties as a *grex* (flock, or mix) of 17 Ohio Corn Belts dents that reproduced one another through sharing pollen and giving each other stories once more.

When the agronomist and I met, I was growing several varieties of open-pollinated corn from New Mexico and actively documenting them through an artwork I call *Corn Morphology* (see Figure 9.5). It is an ongoing archive and time-based imaging process for which I flatbed scan each ear of corn and combine the digital images into video animations; afterwards, I shell the corn, prepare it as food for my family, and save seed to grow the next year. Through this artwork, I am interested in visualizing how corn changes from year to year based on how it expresses itself through the eco-social and how the environment, climate, and human hand might affect it over the course of seasons, years, and generations. When I told him about this project, my family farm history in Ohio, and my interest in corn, he said that he had a gift for me. About two weeks later I received a small bag of 500 seeds from his three years growing the 17 varieties of corn.

Pioneer 17 is an ongoing art project to radicalize these surviving Ohio Corn Belt dents and, in a way, heal the trauma of my own settler-colonial past. I will never be able to erase what my colonial ancestors did, but what if this act of decolonization can germinate a new story and life for this corn? My intentions for this artwork are to plant the corn and allow it to grow, commingle, and thrive within the ecological imaginary of the farm for the rest of my life; to dismantle its colonial bonds by allowing it to grow in a polyculture of many selves, without the capitalist mythos of "more, better, faster," and through transforming our entanglement into a form of deep empathetic caregiving; and to give it a place to share its story through the seeds it makes, the sustenance it shares, and the kernels of creative practice we, as an ecological community, cultivate together.

Since 2016, I have shared *Pioneer 17* through performative public talks.[7] The project also includes a growing archive of still and moving images as I document the transformation of intersubjective forms and stories into a collective power of sympoiesis (see Figure 9.6). The diversity and variation expressed in the corn's color, shape, size, and pattern make these images queer

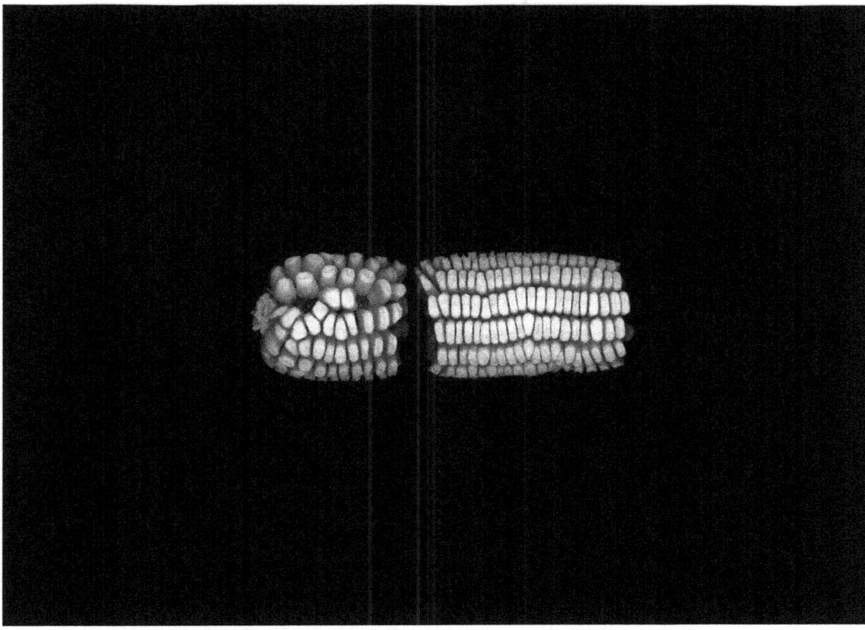

FIGURE 9.6 Jeanette Hart-Mann, *Pioneer 17*, 2020, digital flatbed scan. Image courtesy of the artist.

and tentative as one and many bodies engaged as vegetal creativity and sovereignty. *Pioneer 17* evokes my continued interest in the perceptual space of the lapse, gap, and glitch where the senses can breathe anew and listen to stories never heard (see Figure 9.7). Most important, like all of my work, from farming to art, this project and the process it demands are generative and iterative as they follow these 17 radical evocations of seed along new paths and stories.

Industrial agriculture is one of the most powerful forces driving the destruction of all life on the planet, leading to desertification, biodiversity loss, toxic pollution, food and water insecurity, and the violent exploitation of plants, animals, environments, and people. Technocrats proclaim its virtue and argue that it is the only way to feed people now and in the future, but dependency on destruction as nourishment is an atrocity—and it is the crisis we face. Botanists call corn plants "hopeless monsters" because corn has evolved over the last 10,000 years through a codependency with humans (Fussell 2004, 45). This eco-social relationship relies on farmers to harvest and replant seed every year. What this hopeless sentiment portrays is a worldview that begets the nature–culture divide separating humans from nature and perceiving ecological interrelationships as monstrous. Furthermore, this framework normalizes violent domination which embeds us perpetually in environmental devastation.

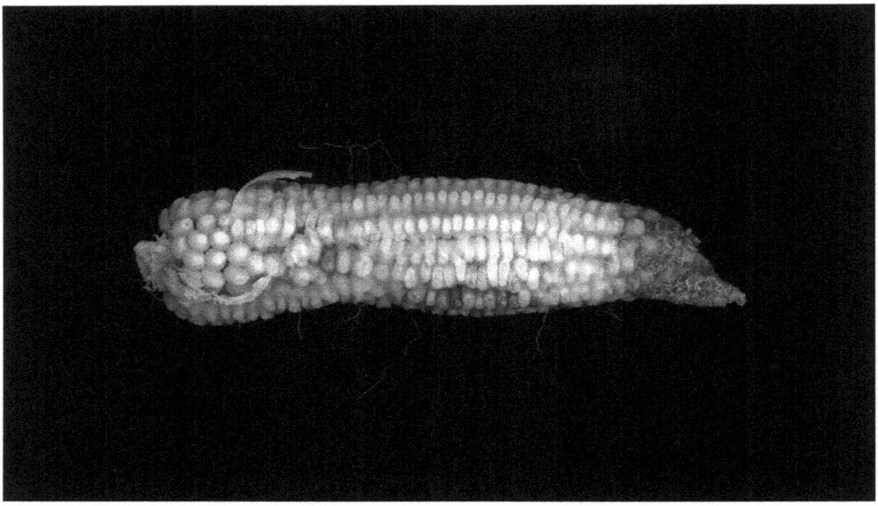

FIGURE 9.7 Jeanette Hart-Mann, *Pioneer 17*, 2020, video still. Image courtesy of the artist.

Instead, we need corn as much as corn needs us. Ecological entanglements create life on Earth. Without complex and diverse eco-social relationships, there would be no seeds, plants, or food. There would be no us, you, or me. My work as an artist and a farmer is rooted in this counterpoint to the mythos of colonial and capitalist domination where stories germinate and grow through ecologies of more-than-human making with. This eco-social storymaking incites curiosity, queerness, creativity, and resistance to cultivate futurities of ecologic healing and hope. Seed stories are powerful reminders that multispecies dependencies are the stories and relations we need to cultivate right now. Seeds are teachers, caregivers, and revolutionaries, but they can only thrive if we are willing to listen, value, share, and reseed their stories. A radical evocation of seed is a call to action to incite multispecies worldmaking through story.

Notes

1 "More-than-human" refers to all ecological selves that exist in the world who are not human. I first learned this term from Navajo medicine man and farmer Larry Emerson from Tse Daa K'aan, Navajo Nation, although the concept is typically associated with the work of American ecologist and philosopher David Abram from his book *The Spell of the Sensuous* (1996), meaning "the sensuous world that surrounds us."

2 This question of intention was first taught to me by Richard Moore, who is an environmental justice community organizer, farmer, and director of Los Jardines Institute in Albuquerque, New Mexico.

3 SeedBroadcast was cofounded with Chrissie Orr in 2011 (https://www.seedbroadcast.org); artist Ruben Olguin joined from 2015 to 2019. SeedBroadcast created "agri-Culture" as a textual concept that is a play on the word agriculture. It suggests ideas

and practices that are often neglected in industrial farming, land policies, and the global food system; that is, that culture(s), daily lives, beliefs, and values are interwoven into the food we eat, the way it is grown, and hence our ecological impact on the world. Culture is, likewise, a catalyst in creating the agricultural system we perpetuate.

4 Community partners and collaborators include Acoma Ancestral Farm Corp, Española Healing Foods Oasis, and Ron Boyd of Mergirl Gardens.

5 In 2012, I disseminated *Letter from a SeedBroadcaster* as a video letter that I sent out to individuals and organizations across the United States. SeedBroadcast played it in the Mobile Seed Story Broadcasting Station during all public events and I self-published it in the book *SeedBroadcast: Logics of Inquiring Diversity.*

6 Not only was corn unwillingly appropriated from the legacy of Indigenous people to become the world's leading crop, but land grant universities were created by the Morrill Act, which stole tribal lands and deeded them to land grant institutions so they could in turn sell the land and fund themselves. See Cailin Hunt (2020).

7 For example, in *Oppression/Pollination/Agency*, SOMA, Mexico City, Mexico, 2016, and in 2020 *Rural Environmentalisms: A Roundtable*, Harwood Museum of Art, Taos, New Mexico, 2020.

References

Abram, David. 1996. *The Spell of the Sensuous: Perception and Language in a More-than-Human World.* New York: Pantheon.

Bennett, Jane. 2010. *Vibrant Matter: A Political Ecology of Things.* Durham, NC: Duke University Press Books.

Capehart, Tom, and Susan Proper. 2019. "Corn Is America's Largest Crop in 2019." USDA. Accessed July 29, 2021. https://www.usda.gov/media/blog/2019/07/29/corn-americas-largest-crop-2019.

Center for Food Safety. n.d. About Genetically Engineered Foods." Accessed February 28, 2022. https://www.centerforfoodsafety.org/issues/311/ge-foods/about-ge-foods.

Freire, Paulo. 2000. *Cultural Action for Freedom*, 2000 edition, with an introduction by Marta Soler-Gallart and Bárbara M. Brizuela. Cambridge, MA: Harvard Educational Review.

Friis, Else Marie, Peter R. Crane, and Kaj R. Pedersen. 2011. *Early Flowers and Angiosperm Evolution*. Cambridge: Cambridge University Press.

Fussell, Betty. 2004. *The Story of Corn.* Albuquerque: The University of New Mexico Press.

Ghosh, Amitav. 2021. *The Nutmeg's Curse: Parables for a Planet in Crisis.* Chicago: University of Chicago Press.

Gliessman, Stephen R. 2015. *Agroecology: The Ecology of Sustainable Food Systems*, 3rd ed. Boca Raton, FL: CRC Press.

Haraway, Donna J. 2016. *Staying with the Trouble: Making Kin in the Chthulucene.* Durham, NC: Duke University Press Books.

Hudson, John C. 1994. *Making the Corn Belt: A Geographical History of Middle-Western Agriculture.* Bloomington: Indiana University Press.

Hunt, Cailin. 2020. "Interactive: See the Lands Sold to Fund Ohio's Land 'Grant' Universities." Pulitzer Center. August 7, 2020. Accessed July 29, 2021. https://pulitzercenter.org/stories/interactive-see-lands-sold-fund-ohios-land-grant-universities.

IPBES. 2019. "Media Release: Nature's Dangerous Decline 'Unprecedented'; Species Extinction Rates 'Accelerating.'" Accessed February 28, 2022. https://www.ipbes.net/news/Media-Release-Global-Assessment.

Kimmerer, Robin Wall. 2015. *Braiding Sweetgrass: Indigenous Wisdom, Scientific Knowledge and the Teachings of Plants.* Minneapolis, MN: Milkweed Editions.

Kohn, Eduardo. 2013. *How Forests Think: Toward an Anthropology beyond the Human.* Berkeley: University of California Press.

Marder, Michael. 2016. "Vegetal Anarchitectures." *The Philosopher's Plant. Los Angeles Review of Books,* June 21, 2016. Accessed February 28, 2022. https://philosoplant.lareviewofbooks.org/?p=164.

Pollan, Michael. 2001. *The Botany of Desire: A Plant's Eye View of the World.* New York: Random House.

Smith, Frank Percy, dir. 1910. *The Birth of a Flower.* Kineto Films. Accessed February 28, 2023. http://www.screenonline.org.uk/film/id/594372/index.html.

UN News. 2010. "Conserving Plant Genetic Diversity Crucial for Future Food Security—UN." United Nations. October 26, 2010. Accessed February 28, 2022. https://news.un.org/en/story/2010/10/357072-conserving-plant-genetic-diversity-crucial-future-food-security-un.

INDEX

For Product Safety Concerns and Information please contact our EU
representative GPSR@taylorandfrancis.com
Taylor & Francis Verlag GmbH, Kaufingerstraße 24, 80331 München, Germany

www.ingramcontent.com/pod-product-compliance
Ingram Content Group UK Ltd.
Pitfield, Milton Keynes, MK11 3LW, UK
UKHW020149200926
475702UK00016B/2620

9781032391342